PRINCIPLES OF HOLOGRAPHY

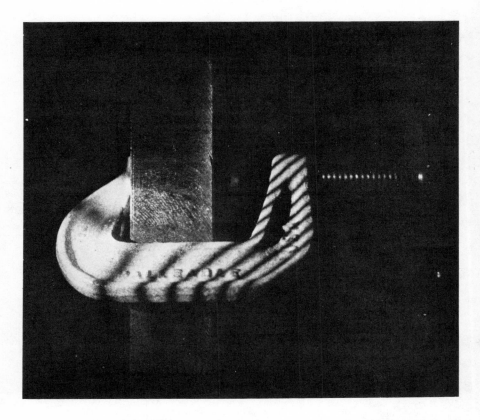

An example of double exposure holographic interferometry. A hologram of the c-clamp is exposed. The light source is then shuttered and the clamp tightened. A hologram of the strained c-clamp is exposed onto the same recording material. After processing, illumination of the doubly exposed hologram results in the reconstruction of two wavefronts—one corresponding to each amount of strain on the c-clamp. These two waves interfere so as to form the fringes pictured here.

Principles of

HOLOGRAPHY

HOWARD M. SMITH

Senior Research Physicist
Research Laboratories
Eastman Kodak Company
Rochester, New York

WILEY—INTERSCIENCE

A Division of John Wiley & Sons

NEW YORK · LONDON · SYDNEY · TORONTO

Library of Congress Catalog Card Number: 69-16129
SBN 471 08340 2
Printed in the United States of America

For My Father and Mother

Preface

The field of holography has advanced rapidly since its renaissance in the period between 1962 and 1964. More than 500 papers and articles have been written on the subject since its inception in 1948—the large majority published since 1964.* This rate of publication has led to duplication, incomplete treatments, erroneous statements, and a careless nomenclature. Within this great volume of published material, however, are several major papers that treat one aspect or another of the science of holography in a definitive manner. Because of this somewhat confusing proliferation of writings on the subject, one of the main purposes of this book is to present a unified and complete treatment under one cover. To do this I have presented with little or no change some of the more important approaches to the various subjects. Of course, the whole idea of off-axis holography must be credited to Emmett Leith and Juris Upatnieks of the University of Michigan.† The work in Section 5.3, which deals with the effects of the recording medium on image resolution, has been borrowed almost entirely from the two papers by Raoul Van Ligten of the American Optical Company.‡ Section 5.4 deals with the third-order aberrations of the holographic process and has been taken virtually unchanged from the paper by R. W. Meier of Xerox.§ Finally, Section 6.1.2 is made up entirely of the work presented in the papers by Kaspar and Lamberts,‖ and Kaspar, Lamberts, and Edgett¶ of Eastman Kodak. The rest of the material has been gleaned from other publications or has been worked out by me, usually in a very simple fashion.

    Some of the terminology and usage associated with this field has become

---

* J. N. Latta, *J. Soc. Motion Picture Television Engrs.*, **77**, 422 (1968).
† E. N. Leith and J. Upatnieks, *J. Opt. Soc. Am.*, **52**, 1123 (1962).
‡ R. F. Van Ligten, *J. Opt. Soc. Am.*, **56**, 1, 1009 (1966).
§ R. W. Meier, *J. Opt. Soc. Am.*, **56**, 219 (1966).
‖ F. G. Kaspar and R. L. Lamberts, *J. Opt. Soc. Am.*, **58**, 970 (1968).
¶ F. G. Kaspar, R. L. Lamberts, and C. D. Edgett, *J. Opt. Soc. Am.*, **58**, 1289 (1968).

inaccurate and inconsistent, due in part, I suppose, to the large volume of papers published each month in the technical journals. I have tried to use a "correct" terminology and I hope that it will be generally accepted. I have never employed the phrase "to reconstruct an image," since in holography it is the wavefront that is being reconstructed and not the image. Also, I have never employed the phrases "to reconstruct the object" or, worse yet, "to reconstruct the hologram." One illuminates the hologram in such a way that a wavefront is reconstructed which can be used to form an image of the object. This is the usage I have tried to maintain. The two images associated with the holographic process are almost universally referred to as the real and virtual images, even though this designation does not specify completely the image to which we refer. The "real" image, for example, may very well be virtual, which can only lead to confusion. I have therefore consistently designated the two holographic images as the *primary* and *conjugate* images.

These are small points, to be sure, but I feel that a careless and loose terminology tends to degrade the field and confuse those not familiar with conventional usage.

I have tried to keep the physics as simple as possible, so that the reader can get a clear insight into what is going on in any given situation. The field of holography is not a complex or intricate one in any case—most things can be understood from a fundamental viewpoint and I have attempted to stress this in the treatment of each topic. The book is written so that any photographic or optical engineer will be able to learn the whys and wherefores of any aspect of holography and thus be able to deal intelligently with any questions that might arise, regardless of whether he is working in the laboratory or making management-level decisions concerning holography. It is assumed that the reader has a knowledge of several aspects of optics, such as modulation transfer functions, spatial frequency, phase, coherence, diffraction, and so on. Some of these concepts are treated only briefly when they are introduced.

Finally, I should warn the reader that some material is repeated from time to time for the sake of completeness. The lensless Fourier transform hologram is described in Chapter 3 and again in Chapter 6. The equation describing the exposure at the hologram phase has been included almost every time a new topic is introduced. The integral specifying the object field at the hologram plane appears several times throughout the text. These things, and possibly some others, have been repeated so that each chapter, and even each topic, can be read and understood almost independently of the others.

Howard M. Smith

Rochester, New York
June, 1968

Contents

PRINCIPLES OF HOLOGRAPHY

1 Historical Introduction

Ever since 1900 man has been able to record and retain as a permanent record almost any scene that his eyes perceived—through the process of photography. The optical lens had been invented and used several centuries before, and the formation of optical images with lenses was well understood by 1900. With the invention of the photographic process the importance of the lens in scientific investigation was greatly enhanced. The fortunate combination of lens and photographic emulsion made possible the charting of stars, planets, and galaxies; the recording of optical spectra; the picturing of minute microscopic specimens; the storage of large amounts of data in the form of small recorded images; and myriad other uses. Because of the vast scope of its scientific importance, the science of photography has advanced steadily over the past 70 or more years; even today new and important uses are being found.

Now science has at its disposal a new method of forming optical images: holography.

Holography is a relatively new process which is similar to photography in some respects but is nonetheless fundamentally different. Because of this fundamental difference, holography and photography will not be competing in the same areas. There are several applications for which holography is more suitable than photography, whereas most of the more important uses for photography remain unchallenged. Further, there are several tasks which can be performed with holography but not at all with conventional photography.

In order to point out the fundamental differences between holography and photography, we should understand in a general way how each works.

Photography basically provides a method of recording the two-dimensional irradiance distribution of an image. Generally speaking, each "scene" consists of a large number of reflecting or radiating points of light. The

1

waves from each of these elementary points all contribute to a complete wave, which we will call the "object" wave. This complex wave is transformed by the optical lens in such a way that it collapses into an image of the radiating object. It is this image which is recorded on the photographic emulsion.

Holography is quite different. With holography, one records not the optically formed image of the object but the *object wave itself*. This wave is recorded in such a way that a subsequent illumination of this record serves to *reconstruct* the original object wave, even in the absence of the original object. A visual observation of this reconstructed wavefront then yields a view of the object or scene which is practically *indiscernible from the original*. It is thus the recording of the object wave itself, rather than an image of the object, which constitutes the basic difference between conventional photography and holography.

A brief description of how the object wave is recorded will be useful before tracing the history of holography. One starts with a single monochromatic beam of light which has originated from a very small source. The requirements that this beam of light be monochromatic and that it originate from a small source together form the condition that the light be *coherent*. The requirement of coherence means that the light should be capable of displaying interference effects that are stable in time. This single beam of light is then split into two components, one of which is directed toward the object or scene; the other is directed to a suitable recording medium, usually a photographic emulsion. The component beam that is directed to the object is scattered, or diffracted, by that object. This scattered wave constitutes the object wave, which is now allowed to fall on the recording medium. The wave that proceeds directly to the recording medium is termed the *reference wave*. Since the object and reference waves are mutually coherent, they will form a stable interference pattern when they meet at the recording medium. This interference pattern is a complex system of fringes—spatial variations of irradiance which are recorded in detail on the photographic emulsion. The microscopic details of the interference pattern are unique to the object wave; different object waves (objects) will produce different interference patterns.

The detailed permanent record of this interference pattern on the photographic emulsion is called the "hologram," from which the word "holography" is derived. This photographic record, or hologram, now consists of a complex distribution of clear and opaque areas corresponding to the recorded interference fringes. When the hologram is illuminated with a beam of light which is similar to the original reference wave used to record the hologram, light will only be transmitted through the clear areas, resulting in a complex transmitted wave. Because of the action of the recorded interference fringes, however, this transmitted wave conveniently divides into three separate components, *one of which exactly duplicates the original object wave*. By viewing this reconstructed wavefront, one sees an exact replica of the original object, even though the object is not present during the reconstruction process. Thus holography is a two-step process by which images can be formed. In the first step a complex interference pattern is recorded and becomes the hologram. In the second step the hologram is illuminated in such a way that part of the transmitted light is an exact replica of the original object wave. The fundamental difference between holography and conventional photography is now quite evident.

This method of optical imagery is not really new. Nearly two decades ago British research scientist Dennis Gabor first conceived of, as he called it, "a new two-step method of optical imagery" [1]. It is only in the past few years, however, that the method has become widely known and used. The modern renaissance in holography had to await the general availability of the laser with the great temporal and spatial coherence of its light, but the really significant contributions to Gabor's original idea were more basic in nature.

The general idea of this two-step imaging process was suggested to Gabor by Bragg's x-ray microscope [2]. Bragg had been able to form the image of a crystal lattice by means of diffraction from the photographically recorded x-ray diffraction pattern of the lattice. The basic idea behind Bragg's method is a double-diffraction process, which is the crux of the holographic process. Image formation by double diffraction becomes clear if we note that the field diffracted by an object can be represented as a Fourier transform of the light distribution at the object [3]. Thus the second diffraction becomes a Fourier transform of the Fourier transform of the object, which is an image of the object itself. This means that diffraction from the hologram will reproduce the object wave, provided that *all the amplitudes and phases of both diffractions are preserved*.

It was just this question of phase preservation that represented the basic limitation to Bragg's method. Since he was able to record only irradiances, phase information was discarded. He was thus limited to applying the method only to a restricted class of objects, such as crystal lattices, for which the absolute phase of the diffracted field could be predicted. It is preserving the phase information, or at least rendering the recording of the phase unimportant, which represents the crux of Gabor's method. Bragg was able to circumvent the phase problem by using a class of objects for which a known phase change occurs between the incident and diffracted radiation. By his use of crystals having a center of symmetry, all the scattered radiation was either in phase or 180° out of phase with the incident radiation. Hence by recording the diffraction pattern photographically and therefore recording only irradiance, *no* phase information was lost and another diffraction of this photograph restored the original object wave.

Buerger [4] was later able to extend Bragg's method to crystals that did not have a center of symmetry but nevertheless had approximately known phase changes. By inserting appropriately placed "phase shifters," he was able to reconstruct the true wave of the original object.

Gabor [1] was able to extend these ideas by reasoning that the phase of the diffracted wave could be determined by comparison with a standard reference wave. He added to the diffracted wave a strong, uniform radiation, the amplitude of which is modulated by the diffracted radiation, provided the two are coherent with respect to one another. A photograph of the resulting modulated diffracted wave constitutes the hologram. Diffraction of radiation from the hologram then gives the second diffraction, resulting in the reconstruction of the original wavefront.

The situation, however, is not quite that simple, as Gabor recognized from the start. The photographic plate records the modulated amplitude of the diffracted wave, but it still does not record its absolute phase. Gabor made the diffracted wave weak with respect to the reference wave, so that the phase of the resultant wave (diffracted plus reference) was always approximately that of the reference wave. Thus the diffracted wave could be considered to have the phase of the reference wave and vary only in amplitude. This approximation results in the production of two waves— the original wave diffracted by the object and a "twin wave" that has the same amplitude, but opposite phase, in relation to the reference wave. Efforts to eliminate the effects of the twin wave constituted a large portion of the early work in wavefront reconstruction, or "holography" [1, 5, 6]. The recent complete elimination of this problem [7] was one of the major reasons for the dramatic and widespread resurgence of interest in holography.

Gabor first demonstrated the feasibility of the holographic method in 1948 [8]. The radiation he used was light. The necessary monochromaticity was obtained by filtering the radiation from a mercury arc; the required spatial coherence was obtained by illuminating pinholes. His primary objective was to increase the resolution of the electron microscope. The theoretical resolution limit of the electron microscope at that time was about 5Å, which was determined by a compromise between diffraction and the spherical aberration of electron objectives. Since at that time it seemed unrealistic that any further correction of the aberrations of electron objectives was feasible, Gabor reasoned that by making the hologram with electrons and reconstructing with light, an aberration-free image could be obtained, provided the aberrations of the light optics precisely matched the aberrations of the electron optics. The procedure he suggested was to put the object in the electron beam, just in front of a reduced image of the electron source. A hologram formed by the electrons diffracted from the object was recorded on a photographic plate some distance beyond the object. The hologram was then scaled up optically in the ratio of the light wavelength to the electron wavelength and illuminated with a light wave with the same aberration as the electron wave, scaled in the same ratio. Theoretically, then, the object was visible through the hologram in the original position and magnified by this same ratio of light to electron wavelength, or about 100,000X.

The method did not succeed because of certain technical difficulties, such as mechanical and electrical stability. Haine and Dyson subsequently suggested an improved arrangement which increased the usable field, thus relaxing the electrical stability requirements [9]. By this method, the "transmission method," lenses were used between the object and the hologram to magnify the diffraction pattern. This increased the effective resolution of the photographic plate. The required apparatus was essentially identical with the classical electron microcsope, thus facilitating the location of the object.

Efforts by Haine and Mulvey to implement this method were again frustrated by practical limitations [10]. They managed to obtain diffraction resolutions of about 6Å by eliminating as many of these problems as possible but could not go further because of difficulty in holding the specimen stationary in the electron beam. It was necessary to hold the object stationary in relation to the objective lens to within a few angstrom units during exposure, a period of several minutes. Further difficulties were encountered in the optical reconstruction stage.

The most serious problem in reconstruction was the disturbance created by the twin wave. Because of the uncertainty of π in recording the phase in the hologram, there are two possible objects giving rise to the same exposure distribution in the hologram. One of these is the original object, the other a virtual object located symmetrically behind the source. Upon reconstruction, waves from both objects are formed. Therefore, in viewing the image of the real object, one has to look through an out-of-focus background image of the virtual object, a most annoying disturbance. Gabor [1] noted that if a condenser system is used to form a reduced image of the source, the twin image will be severely affected by the aberrations of that system. Thus it will appear blurred, whereas the image of the original object will appear sharply defined, the aberrations having been compensated for in this image during reconstruction. Gabor [11] also suggested a method whereby an obscuring mask is placed at a suitable formed image of the point source during reconstruction. Thus most of the wave containing the information of the real object is passed, but most of the background is suppressed. Bragg and Rogers [5] suggested that since the reconstruction is really an in-focus image of the object plus a hologram of the conjugate object, a second hologram can be made of the conjugate object and subtracted from the original. The subtraction is performed by placing the second hologram in contact with the original image. The high-transmission portions of the secondary hologram fall on the high-density regions of the unwanted image and vice-versa, so that the background becomes uniform. The great precision required to register the two prevented complete success, but the effects of the twin image were reduced [5, 6].

Paralleling these early attempts to utilize holography for the improvement of electron microscopy were efforts by several workers to produce x-ray holograms. El Sum [6] produced an artificial x-ray hologram of a thin wire by photographing a published picture of the x-ray diffraction fringes of the wire. He managed to obtain a reasonable reconstruction, using light from this hologram, proving at least the feasibility of x-ray holography. Baez [12] did a theoretical study of the problems and also concluded that holograms and reconstructions with x-rays are feasible. He does note, however, that because of film resolution and source size limitations, useful resolution might be achieved with visible light. Further work on x-ray holography is still awaiting a small, monochromatic source of x-rays.

Aside from efforts by Rogers [13, 14] and Kirkpatrick and El Sum [15] to provide more satisfying conceptual explanations, the subject of holography lay dormant for almost a decade. Brief explanations of the principle published in a few optical textbooks [16, 17] represented about all of the published work on the subject for this long period of time. The most serious limitations of the method, which led to interest dying out, were the lack of an intense, coherent source in either the x-ray or optical region of the spectrum and the disturbing presence of the twin wave. Exposure times on the order of one hour were not unusual [6], and resolution in the reconstruction never reached theoretical predictions.

Interest began reviving in the field when Leith and Upatnieks [7] demonstrated a method for the complete elimination of the twin wave by a fairly simple means. Describing the holographic process from a communication-theory viewpoint, they realized that if the signal information (wavefront diffracted from the object) could be put on a carrier frequency (off-axis reference wave), the two reconstructed waves would then represent the sidebands of the process and be physically separated from each other. From an optical viewpoint, if the wave diffracted from the object is made to interfere with a reference wave which is off-axis, rather than in line, the hologram will be a gratinglike structure. Reconstruction will yield two waves representing the two first orders of the grating. One of these waves is the same as the original wave from the object; the other is the unwanted twin wave. Thus a physical separation in space of the two waves is achieved and the disturbing effects of the twin waves are eliminated. Gabor [1] noted that this turn of events would probably occur when he said, ". . . it is very likely that in light optics, where beam splitters are available, methods can be found for providing the coherent background which will allow better separation of object planes and more effective elimination of the effects of the 'twin wave' than the simple arrangements which have been investigated." Baez [12] came very close to introducing the off-axis concept in 1952 when he noted that "in an analogous way a diffraction grating forms a virtual image of a source," while explaining how a hologram forms an image by diffraction.

Thus Leith and Upatnieks' idea revived interest in holography. Many people began taking note and trying a few simple experiments. The "twin image" would no longer be termed the "unwanted image" but would prove to be a useful adjunct of the holographic process. Although the conception of an off-axis reference wave was a definite advance, there were still problems with dust and imperfections in the optical components. The slightest speck of dust on the lenses or mirrors gave rise to its own hologram, reducing the effective aperture of the hologram and reconstructing itself as noise. El Sum [6] and others took great pains to remove this source of noise by rotating as many of the components as possible, thus smearing out the holograms of the nonstationary dust particles. This method was successful to some degree but extremely impractical. Later developments would eliminate this problem also.

Another advantage of the new off-axis reference beam method is elimination of critical film processing. The original method required that the hologram be processed as closely as possible to a gamma (contrast) of two for linear transfer of exposure to amplitude transmission. In the Gabor technique any nonlinearities in the transmission-exposure transfer resulted in decreased image contrast because of the background light level. In the new technique any nonlinearities of the recording medium result mainly in higher diffraction orders. These higher orders are diffracted at angles larger than the first-order wave; thus nonlinear recording has little effect on the desired image. Curiously enough, processing the film to a high contrast with this new method, thereby increasing the nonlinearity of the recording, is actually somewhat beneficial. A high-contrast hologram results in a brighter image but with little disturbing effect from the higher order terms.

The off-axis reference beam method also results in the elimination of the effects of self-interference between different points of the object. In the earlier method this self-interference resulted in a veiling glare around the image. In the new method this noise term can be avoided completely.

Finally, the new method makes reconstruction possible for objects that do not transmit a large portion of the incident wave and also for continuous tone objects. In the earlier technique it was necessary that the major portion of the light passing the object not be diffracted. This undiffracted light is then only slightly modulated by the light diffracted from the object. In this way the loss of phase information in the recording process is rendered negligible; the resultant phase of the total disturbance at the hologram recording plane is almost that of the background wave. When the hologram is illuminated with the background wave alone the phase of the original total disturbance is approximated and a recognizable reconstruction results. In the new method the phase of the total disturbance at the hologram is recorded as a phase modulation of a gratinglike fringe pattern. This permits reconstructions of a wholly new class of objects which do not transmit a large portion of the incident wave, such as transparent letters on an opaque background and continuous-tone objects.

Thus the new method introduced by Leith and Upatnieks eliminated many of the annoying features of the original method, but the real renaissance of holography had to await two other important advances which were not long in coming.

About the same time (1962) that Leith and Upatnieks were introducing the off-axis reference beam method, people were beginning to make and use a radically different light source that would prove to be eminently suitable for holography. Thus the invention of the gas laser coincided nicely with the revival of interest in holography. The laser is capable of producing very intense monochromatic radiation in regions of the spectrum that can be recorded photographically. Because of the highly coherent nature of the light from a laser it can be focused down to an arbitrarily small spot; hence source size no longer limits the attainable resolution in a holographic image. The monochromaticity of the laser allows for full utilization of the off-axis recording scheme, since now many more interference orders (fringes) can be recorded. This yields much higher resolutions than had previously been obtained. Also there are no longer any restrictions on the size of the object to be used; holograms can now be made of very large objects.

The advent of the gas laser made possible still another important advance, again introduced by Leith and Upatnieks. In 1964 they introduced the concept of diffuse illumination holography [18]. Before this the only holograms that had been made were of thin transparent objects. Holograms of these kinds of objects often consisted of nearly recognizable shadowgrams of the object. Thus a small region of the hologram would bear almost a one-to-one correspondence with a small region of the object. Viewing the images formed with this type of hologram required some additional optical components, since an observer viewing a specularly illuminated transparency will see, for the most part, only that portion of the transparency which lies on a line between the light source and his eye pupil. Hence without optical aids, only a small portion of the image can be viewed at one time. On the other hand, if the transparency is illuminated diffusely, it can be viewed in its entirety with the eye in one location. This, then, is the idea that Leith and Upatnieks introduced into holography in 1964. By placing a diffuser, such as an opal glass, behind the object, a hologram is formed of both the diffuser and object. In this way it is possible to view the image formed from the hologram by merely looking through the hologram as through a window. The ability to view the images formed from this type of hologram in this way is not the only advantage. There are several even more interesting.

Because the object is diffusely illuminated there are no longer any recognizable shadowgrams of the object on the hologram. The light scattered from each point of the object spreads out so as to cover the entire photographic plate. This means that there is no longer a one-to-one correspondence between a portion of the hologram and a region of the object. The information about any single object point is recorded over the whole plate. Therefore only a small portion of the hologram is required to form an image of the whole object. In fact, in viewing the hologram with the unaided eye, only a portion of it roughly the size of the eye pupil is actually used. If the hologram were to be broken, scratched, torn, or damaged, it would still be possible to form an image of the complete object, although some resolution would be lost.

A further advantage of holograms of diffusely illuminated objects is that dirt or scratches on the mirrors, beamsplitters, and/or lenses used in making the hologram no longer represent the problem they did in the earlier types. Most of these imperfections are, to a great extent, smeared out over the whole plate and thus have a negligibly small effect on the reconstruction.

Perhaps the single most important aspect of this technique is the ability to record holograms of diffusely reflecting, three-dimensional objects. It had been recognized from the beginning by Gabor that a hologram of a three-dimensional object should be capable of forming a three-dimensional image [1]. El Sum [6] and Rogers [14] managed to make holograms of objects in depth, even with the limited coherence lengths of the light sources at their disposal. The gas laser and the diffuse illumination concept made possible the formation of truly striking three-dimensional images. Since the object is diffusely illuminated or diffusely reflecting, light from a large range of perspectives reaches the photographic plate. An observer viewing the image formed by this hologram can move his head and see around foreground objects, just as if he were viewing the original object; he sees a truly three-dimensional image. He must refocus his eyes, depending on whether he is viewing a near or far object point.

The new diffuse-illumination holograms also lend themselves very nicely to the superposition of more than one hologram on a single photographic plate. Rogers [14] had made two holograms on a single plate by double exposure, but they were of two objects suitably situated so that the hologram of one did not obliterate too great a portion of the hologram of the other. Suitable positioning of the objects was necessary with the Gabor-type hologram since the information about an object was fairly well localized in the region surrounding the geometrical shadow of the object. Since this is not the case with diffuse-illumination holograms, it is a simple matter to make multiple holograms on a single plate and still obtain high-quality reconstructions. There is still a limitation on object position, however. To prevent the various images from each hologram from falling on top of one another, the hologram of each object should be recorded with a different angle between the object beam and reference beam. In this way the reconstruction of each object wavefront will be traveling in a different direction and hence will be separated in space. Each reconstruction can thus be viewed separately.

This concept led Leith and Upatnieks to propose a method of multicolor wavefront reconstruction [18]. By illuminating a colored object with coherent light in each of the three primary colors, each with its own reference beam, three holograms will be recorded. Reconstruction is accomplished by reilluminating with the three reference beams; a full-color object wavefront results. The scheme has been demonstrated, but better methods have evolved. They will be discussed in Chapter 7.

At about the same time that Leith and Upatnieks were advancing the field of holography at a great rate, Denisyuk [19] proposed an idea that proved to be of fundamental significance. He suggested that the wavefront from an object traveling in one direction be made to interfere with a coherent reference beam *traveling in the opposite direction* in a three-dimensional recording medium. In this way a standing wave pattern is set up in the recording medium which is uniquely related to the object wavefront. The medium therefore records a series of surfaces separated by one-half the recording wavelength in the medium. These surfaces, under appropriate conditions, are just the antinodal surfaces of the object wavefront. In this way the actual wavefronts of the object beam are recorded. The recording medium is processed so that a change in the dielectric constant occurs where the exposure is high. Reillumination with the reference wave alone yields a reflected wave that is an exact replica of the object wave. Denisyuk's idea was thus a fruitful combination of Lippmann's [20] color photographic process and the hologram method of Gabor [1]. This idea was later investigated in detail by van Heerden [21] and has proved to be an important aspect of modern holography.

The holographic method differs significantly from the conventional photographic process in several basic respects and has distinct advantages in many areas. The most obvious advantage of holography is the ability to store enough information about the object in the hologram to produce a true three-dimensional image, complete with parallax and large depth of focus. There has been a great deal of work done in the attempt to produce three-dimensional images using conventional photographic techniques. These methods have been only partially successful because of the limited depth of field and restricted viewing conditions. An observer viewing a stereo pair, for example, cannot move his head from side to side and look behind foreground objects as he can with a hologram. The lenticular-type three-dimensional photograph allows limited parallax but has a rather severe depth limitation. The hologram, on the other hand, has a field of view that is limited in general only by the resolution of the recording medium. The depth of field recorded in a hologram is limited only by source bandwidth. Thus if a hologram is made of a three-dimensional object, it is equivalent to many conventional photographs, each taken from a different point of view and each focused at a different depth. Subsequent viewing of the hologram image at different depths requires only a refocusing of the viewing system. Hence it is fair to say that one hologram is worth a thousand pictures!

The quality of a holographic image is less sensitive to the characteristics of the recording medium than is the quality of a photograph. Holograms made on high-contrast material reproduce tonal variations of the object over a wide range. Nonlinear recording has only a small effect on the final image. Also, imperfections in the emulsion, such as scratches, have very little effect on the final image. Indeed a modern hologram is so redundant that only a small fraction of the holographic record is necessary to form a complete image.

Because of these basic differences between holography and conventional photography, many interesting and novel applications have been proposed. Few of these have been put to commercial use as yet, but the field is still young.

REFERENCES

[1] D. Gabor, *Proc. Roy. Soc.* (London), Ser. A, **197**, 454 (1949).
[2] W. L. Bragg, *Nature*, **149**, 470 (1942).
[3] F. Zernike, *Ned. Tijdschr. Natuurk.*, **9**, 357 (1942).
[4] M. J. Beurger, *J. Appl. Phys.*, **21**, 909 (1950).
[5] W. L. Bragg and G. L. Rogers, *Nature*, **167**, 190 (1951).
[6] H. M. A. El Sum, *Reconstructed Wavefront Microscopy*, Ph.D. Thesis, Stanford Univ., November, 1952.

[7] E. N. Leith and J. Upatnieks, *J. Opt. Soc. Am.*, **52,** 1123 (1962).

[8] D. Gabor, *Nature*, **161,** 777 (1948).

[9] M. E. Haine and J. Dyson, *Nature*, **166,** 315 (1950).

[10] M. E. Haine and T. Mulvey, *J. Opt. Soc. Am.*, **42,** 763 (1952).

[11] D. Gabor, *Proc. Phys. Soc.* (London), **B64,** 449 (1951).

[12] A. V. Baez, *J. Opt. Soc. Am.*, **42,** 756 (1952).

[13] G. L. Rogers, *Nature*, **166,** 237 (1950).

[14] G. L. Rogers, *Proc. Roy. Soc.* (Edinburgh), **A63,** 14 (1952).

[15] P. Kirkpatrick and H. M. A. El Sum, *J. Opt. Soc. Am.*, **46,** 825 (1956).

[16] R. S. Longhurst, *Geometrical and Physical Optics*, Longmans, Green and Co., Ltd., London, 1957, p. 301.

[17] M. Born and E. Wolf, *Principles of Optics*, The Macmillan Co., New York, 1959, p. 453.

[18] E. N. Leith and J. Upatnieks, *J. Opt. Soc. Am.*, **54,** 1295 (1964).

[19] Y. N. Denisyuk, *Soviet Phys.*—Doklady, **7,** 543 (1962).

[20] G. Lippmann, *J. Phys. Radium*, **3,** 97 (1948).

[21] P. J. van Heerden, *Appl. Opt.*, **7,** 393 (1963).

2 Basic Arrangements for Holography

2.0 INTRODUCTION

Our discussion of holography begins with a description of the general arrangements currently used for recording and reconstructing. The discussion is restricted to the off-axis type of hologram first described by Leith and Upatnieks [1]. If this scheme is used, the object and reference beams are coincident at the recording medium, arriving from substantially different directions. This is achieved in practice by placing the object laterally some distance away from the source of the reference beam. The object is, of course, illuminated with a beam of light from the same source that provides the reference beam. As described in Chapter 1, the recording medium records the two-beam interference pattern; the precise details of the pattern are unique to the object used. This record is now called the hologram of the object and when it is illuminated with a single beam of light similar to the original reference wave, the hologram diffracts the light in such a way as to reconstruct the object wave.

2.1 BASIC DESCRIPTION OF HOLOGRAPHY

Conceptually, the simplest form of an off-axis hologram is one for which the object is just a single, infinitely distant point, so that the object wave at the recording medium is a plane wave (Fig. 2.1a). If the reference wave is also plane and incident on the recording medium at an angle to the object wave, the hologram will consist of a series of Young's interference fringes. These recorded fringes are equally spaced straight lines running perpendicular to the plane of the figure. Since the hologram con-

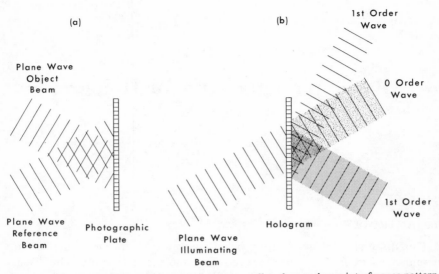

Fig. 2.1   The basic plane wave hologram. (*a*) Recording the two-beam interference pattern. (*b*) The diffraction of a plane wave from the recorded interference pattern.

sists of a series of alternately clear and opaque strips, it is in the form of a diffraction grating. When the hologram is illuminated with a plane wave (Fig. 2.1*b*), the transmitted light consists of a zero-order wave traveling in the direction of the illuminating wave, plus two first-order waves. The higher diffracted orders are generally missing or very weak since the irradiance distribution of a two-beam interference pattern is sinusoidal. As long as the recording is sensibly linear (irradiance proportional to the final amplitude transmittance), the hologram will be a diffraction grating varying sinusoidally in transmittance, and only the first diffracted orders will be observed. One of these first-order waves will be traveling in the same direction as the object wave; this wave has been reconstructed. The other diffracted wave is traveling in a third direction and is called the conjugate wave. This is the twin wave that gave rise to so much trouble in Gabor's early scheme, since it was traveling in the same direction as both the directly transmitted (zero-order) wave and the object wave. With the off-axis method all three of these waves are spatially separated.

The recording of a hologram of a more complicated object can be described with the aid of Fig. 2.2. Let O be a monochromatic wave from the object incident on the recording medium H, and let R be a wave coherent with O. The wave O contains information about the object since the object has uniquely determined the amplitude and phase of O. The object can be thin and transmitting, such as a transparency, or it can be opaque and diffusely reflecting. To make a hologram, then, we must record the incident wave field on a photosensitive medium (usually a photographic emulsion) in such a way that the entire wave can later be reconstructed. To do this, it is necessary to record both the amplitude and phase of the wave O, which is done with the aid of the reference beam R. The total field on H is $O + R$. A square-law recording medium, such as a photographic emulsion, will respond to the irradiance of the light $|O + R|^2$. We now assume that, after processing, the hologram possesses a certain amplitude transmittance $t(x)$ which may be expressed as a function of the exposure

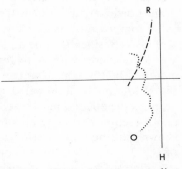

Fig. 2.2 Illustrating the recording of a general wavefront. The curves O and R are schematic representations of the object and reference wavefronts, respectively.

$$t(x) = f[E(x)].$$

By expanding this into a Taylor's series about the average exposure E_o and retaining only the first two terms (see Section 6.1.2 for more detail), we can write

$$t(x) = f(E_o) + \beta E(x). \tag{2.1}$$

For our present purposes, we will simply ignore the constant term $f(E_o)$ and write the amplitude transmittance as

$$t(x) = \beta E(x) = \beta |O + R|^2$$
$$= \beta(|O|^2 + |R|^2 + OR^* + O^*R), \tag{2.2}$$

where the * denotes a complex conjugate. Let us assume that the hologram with this amplitude transmittance is illuminated with the reference wave R alone. The transmitted field at the hologram is then

$$\Psi(x) = R(x) \cdot t(x) = \beta[R|O|^2 + R|R|^2 + |R|^2 O + R^2 O^*]. \tag{2.3}$$

If the reference wave R is now sufficiently uniform so that $|R|^2$ is approximately constant across the hologram, then the third term of Eq. 2.3 is $\beta|R|^2 O = \text{const} \times O$. This term, then, represents a wave identical to the object wave O. This wave has all of the properties of the original wave and can form an image of the object. That this wave is separated from the rest can be seen most clearly by analogy with the diffraction grating hologram previously described. If we consider the complex wave O of Fig. 2.2 to be composed of many plane waves of different amplitudes and directions and also consider R to be a plane wave, then each of the component plane waves interferes with R to form a grating, as in Fig. 2.1. The hologram can then be thought of as consisting of a large number of gratings. When the hologram is illuminated by the reference beam alone, each of these gratings produces a first-order diffracted plane wave of such an amplitude and direction that they all sum to the composite wave O. It can also be shown that the other first-order diffracted wave arises from the term $\beta O^* R^2$ of Eq. 2.3 and that the zero-order wave is expressed by the first two terms $\beta R(|O|^2 + |R|^2)$ of (2.3). In this way we see that the object wave O is clearly separated from the others and may be viewed independently.

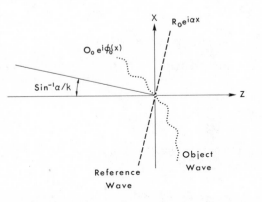

Fig. 2.3 Mathematical expressions describing the object and reference wavefronts.

Another way to explain this separation, using the work of Leith and Upatnieks, is by analogy with the idea of a carrier frequency [1]. Assume that the reference wave R is a plane wave which is incident on the hologram plane at an angle to the object wave O. This is shown in Fig. 2.3. Let the reference wave be written $R_o e^{i\alpha x}$, where R_o is a constant, and the linear phase shift αx indicates that the wave is incident at some angle $\varphi = \sin^{-1}(\alpha/k)$, where k is the wave number of the light $2\pi/\lambda$. If we write the object wave as $O_o e^{i\varphi_o}$ so that O_o and R_o are real numbers, Eq. 2.2 becomes

$$t(x) = \beta[O_o^2 + R_o^2 + 2O_o R_o \cos(\alpha x - \varphi_o)]. \tag{2.4}$$

The information $O_o e^{i\varphi_o}$ in the object wave has been transferred to a spatial carrier wave $\cos(\alpha x)$. The amplitude O_o amplitude modulates this carrier and the phase φ_o phase modulates the carrier. By adding the object wave to a spatial carrier frequency we have been able to achieve the desired separation.

The degree of separation of the various terms of Eq. 2.4 depends on their spatial frequency content. Equation 2.4 describes in one dimension the spatial variation of amplitude transmittance $t(x)$. We can think of $t(x)$ as being composed of a large number of sinusoidally varying transmission gratings, each with a different spatial frequency and phase angle (Fig. 2.4).

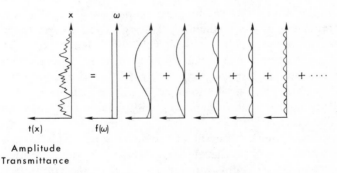

$t(x)$ $\qquad$ $f(\omega)$

Amplitude
Transmittance

Fig. 2.4 Schematic representation of the Fourier decomposition of the transmission function $t(x)$ into a sum of elementary diffraction gratings $f(\omega)$.

The spatial frequencies of these composite gratings vary from 0 to some maximum radian spatial frequency $\omega_{\max}$. This is the principle of Fourier decomposition. Formally, we can write

$$t(x) = \int_0^\infty f(\omega)e^{-i\omega x}\, d\omega \tag{2.5}$$

so that $f(\omega_o)e^{-i\omega_o x}$ describes a single grating of radian spatial frequency ω_o and amplitude $f(\omega_o)$. When a plane wave of light is incident on the plane (Fig. 2.5), each elementary grating diffracts the light in a direction

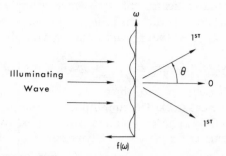

Fig. 2.5   Diffraction from an elementary grating.

determined by the spatial frequency of the grating, based on the grating equation

$$\sin \theta = \frac{\lambda \omega}{2\pi} \tag{2.6}$$

where λ is the wavelength of the incident plane wave and ω is the radian spatial frequency of the elementary grating. The superposition of all of the diffracted waves yields the total field diffracted by $t(x)$. The largest angle θ (Eq. 2.6) at which any light will be diffracted depends on the largest spatial frequency ω_{max} contained in $t(x)$. If $t(x)$ varies only slowly, then ω_{max} will be relatively small and hence θ will remain small. If $t(x)$ varies rapidly from point-to-point, then ω_{max} will be large and some light will be diffracted at large angles.

The spatial frequency content of the various terms of Eq. 2.4 depends on how each term was produced. In most cases, R_o is independent of x, so that there will be no point-to-point variation in the transmittance. For this term $\omega_{max} = 0$ and no light will be diffracted away from the directly transmitted beam because of this term. On the other hand, O_o is usually a function of x, $O_o(x)$. This occurs because of self-interference between different object points, such as points a and b of Fig. 2.6. These two points will interfere and produce an interference pattern with radian spatial frequency

$$\omega = 2k \sin\left(\frac{\varphi}{2}\right) \cos\left(\theta_R - \frac{\varphi}{2}\right); \tag{2.7}$$

the angles are defined in Fig. 2.6. Thus the larger the angle φ is, the greater the spatial frequency of the resultant grating will be. Hence ω_{max} for the $O_o{}^2(x)$ term will depend on the angular extent of the object: the larger the angular extent of the object, the greater the spatial frequency content of

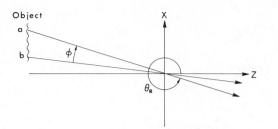

Fig. 2.6 Illustrating the origin of the transmittance fluctuations caused by the object alone. The interference of light from object points such as *a* and *b* produce low spatial frequency exposure fluctuations at the hologram plane.

$O_o(x)$. The light diffracted from the elementary gratings comprising $O_o^2(x)$ is termed *flare light*, since ω_{max} is usually relatively small and the diffracted light caused by this term generally extends only to small angles from the directly transmitted light.

The image-producing third term of Eq. 2.4 has all of the spatial frequency components associated with $O_o(x)$ and also that of the *carrier wave*, $\cos(\alpha x)$. If α of Fig. 2.3 is made large enough the images will be formed at large angles from the directly transmitted and flare light, since ω_{max} for this term will be $k\alpha$ *plus* the ω_{max} of the object wave. In this way we can achieve the desired separation of the image and noise terms of (2.4). A schematic representation of the spectrum of the various terms of (2.4) is shown in Fig. 2.7.

Any nonlinearity in the recording process gives rise to higher diffracted orders, but these will generally be weak and not overlap the desired waves. The elimination of the problems of nonlinear recording is one of the major advantages of off-axis holograms.

The third term of (2.4) leads to two images; these are separated from

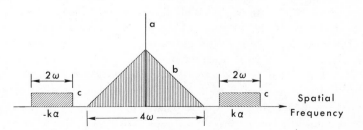

Fig. 2.7 A schematic representation of the spatial frequency spectrum of a hologram (after Leith and Upatnieks [1]).

each other, as well as from the flare light. To show this more clearly, we write (2.4) in the form

$$t(x) = \beta[O_o^2 + R_o^2 + O_oR_oe^{i(\alpha x - \varphi_o)} + O_oR_oe^{-i(\alpha x - \varphi_o)}], \qquad (2.8)$$

so that the carrier wave is expressed by the term $e^{i\alpha x}$. If we now illuminate the hologram with a wave in the same direction as the reference wave, say $C(x) = C_oe^{i\alpha x}$, the third and fourth terms of (2.8) yield the two first-order waves

$$\beta C_oO_oR_oe^{2i\alpha x - i\varphi_o} \qquad (2.9)$$

and

$$\beta C_oO_oR_oe^{i\varphi_o}. \qquad (2.10)$$

The wave (2.10) is identical to the original object wave except for the un-important multiplicative constants. This wave is traveling in the same direction as the original object wave, but the wave described by (2.9) is traveling in a different direction. The direction of this wave is determined by the phase factor $e^{2i\alpha x}$. To see why, assume for the moment that φ_o is constant, that is, does not depend on x. Then the exponent $2\alpha x$ describes the phase of the field in the x plane. The wavefront corresponding to such a phase in the x plane is a *plane* wavefront tilted at an angle $\sin^{-1}(2\alpha/k)$ to the plane as shown in Fig. 2.8a. The phase $\varphi_o(x)$ might correspond to a wavefront as shown in Fig. 2.8b, for example, so that the wavefront corre-sponding to the phase $2\alpha x - \varphi_o(x)$ might appear as in Fig. 2.8c. The effect of the carrier $2\alpha x$ has been to change the general direction of propagation of the wave $e^{i\varphi_o(x)}$. Hence the two waves described by (2.9) and (2.10) are traveling in different directions and will eventually separate.

Since the phase function $\varphi_o(x)$ is *subtracted* from the carrier $2\alpha x$ in (2.9), the wave is said to be *phase conjugate* to the wave $e^{i\varphi_o(x)}$. The conjugate wave produces an image lying on the opposite side of the hologram from the illuminating source, whereas the wave $e^{i\varphi_o(x)}$ produces an image in the same location as the original object. The fact that (2.9) describes a wave which is phase conjugate to the original object wave means that the wave-front is inverted. If the original object wave were a *diverging* spherical wave, for example, (2.9) would describe a *converging* spherical wave. Usually, the object wave is divergent, so $e^{i\varphi_o(x)}$ leads to a virtual image of the object, whereas $e^{-i\varphi_o(x)}$ describes a converging wave that forms a real image of the object. Therefore the third term of (2.8) is often referred to as the real image term and the fourth term as the virtual image term. These designations, however, should not imply that these terms always lead to either a real or virtual image. There are many possible arrange-ments wherein they do not, so this designation is merely convenient for

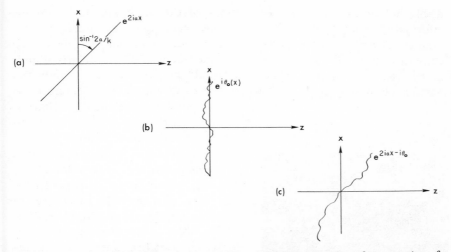

Fig. 2.8 Illustrating the effect of a linear phase shift on the direction of propagation of a wave. In (*a*), a wave that is described by a phase which varies linearly with the coordinate is tilted in relation to that coordinate axis and so is propagating at an angle to it; if this linear phase shift is added to a general wave (*b*), the direction of propagation of this wave is altered (*c*).

labeling the various terms. We will call the image resulting from the third term the *primary* image and that from the fourth term the *conjugate* image.

So far we have not mentioned much about the types of object suitable for holography. Since the original Gabor scheme of in-line holography required a strong background wave to reduce the effects of the twin image, the only suitable objects were transparencies such as opaque lettering on a clear background. Large clear areas were necessary to produce the strong coherent background. Thin transparencies were required because of the limited coherence lengths of thermal sources. With the off-axis scheme a reference wave can be produced which is independent of the object wave (but coherent with it) so that it is possible to produce holograms of clear letters on an opaque background, for example. Such a hologram and an image formed with the reconstructed wave is shown in Figs. 2.9*a* and *b*. Note that because of the essentially specular nature of the object, one can almost tell what the object was from the hologram.

Leith and Upatnieks have demonstrated that holograms can be constructed with diffuse illumination of the object, which has some very interesting advantages. Diffusing the light that illuminates the object effectively causes each point of the object to radiate a spherical wave, hence the information concerning each point of the object is spread out over the whole hologram. Therefore each point of the hologram now contains information about the whole object. The really striking feature of holograms made of diffuse objects is that the reconstruction can be viewed without optical aids. In viewing the reconstruction of a specularly illuminated transparency, for example, it would be possible to see only the portion of the object through which the small cone of rays entering the eye pupil had passed (Fig. 2.10a). If the transparency is diffusely illuminated, however, the viewer sees the whole object at once, since light is passing through each point of the object in all directions (Fig. 2.10b).

It is also possible, of course, to make holograms of diffusely reflecting objects. It is this type of object which gives rise to truly striking reconstructions. The object wave is viewed through the hologram as if it were a window and the objects appear behind the window just as in the original scene. The observer may look from above, from underneath, or from any perspective allowed by the window. He may look around foreground objects to see background objects or, in short, he may view the complete scene from a wide range of perspectives.

Figure 2.11a is a hologram of a diffusely illuminated object scene; Figs. 2.11b, c, and d are views of the reconstruction from three different perspectives. Note that the hologram of Fig. 2.11a is quite different in appearance from that of Fig. 2.9a. There are no discernible Fresnel diffraction patterns from the object. The hologram has a uniformly granular appearance. This is the speckle pattern familiar to anyone who has viewed

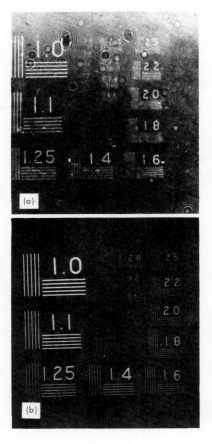

Fig. 2.9 Nondiffuse hologram. (a) Actual hologram of a transparency (clear bars on opaque background) which was specularly illuminated. (b) Image formed with the hologram of (a).

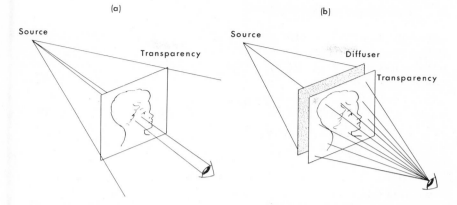

Fig. 2.10 Demonstrating the difficulty in viewing a specularly illuminated transparency compared with a diffusely illuminated one. (*a*) The only portion of the transparency which is visible is that small section intersected by the cone of rays entering the eye. (*b*) Light from all parts of the transparency reach the eye.

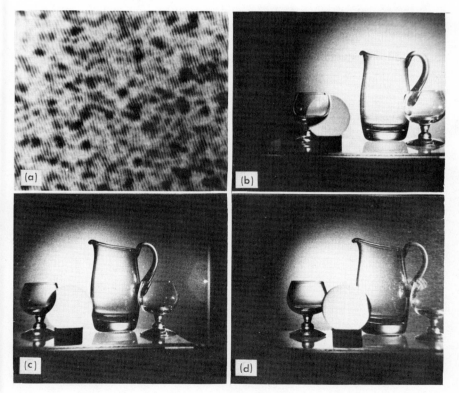

Fig. 2.11 Hologram of a diffusely illuminated object. (*a*) The actual hologram. (*b*), (*c*), (*d*) Three perspectives of the resulting image.

laser light after it has been reflected from or transmitted through a diffuse surface. It is the recording of the $O^2(x)$ term previously discussed. Superposed on this is the fine-line, grating-like structure characteristic of off-axis holograms.

2.2   ARRANGEMENTS FOR RECORDING PLANE HOLOGRAMS

Figure 2.12 illustrates the recording of a hologram and the subsequent reconstruction. As shown in Fig. 2.12*a* the laser beam is first expanded. It is then divided by means of a mirror which directs part of the beam directly onto the photographic plate; the rest of the light reflects from the object. The hologram is the recording of the interference pattern formed by these two beams. After processing, the hologram plate may be replaced in its original position (Fig. 2.12*b*) and the object removed. The light that is diffracted by the hologram forms, in part, the same wavefront that was originally reflected from the object. A viewer looking through the hologram will see an undistorted view of the object, just as if it were still present.

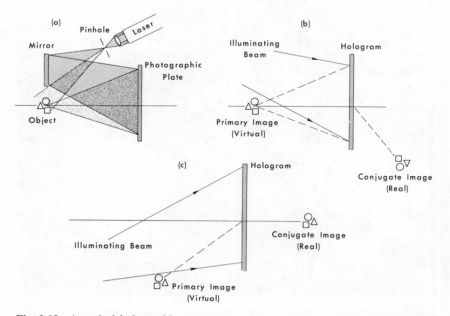

Fig. 2.12   A typical holographic arrangement. (*a*) Recording the hologram. (*b*) Reconstructing the primary object wave. (*c*) Reconstructing an undistorted conjugate wave.

In addition to this virtual, or primary image, a real, or conjugate image will be formed on the observer's side of the hologram. This image will appear unsharp and highly distorted. A distortion-free real image can be formed, however, by changing the position of the illuminating beam so that it appears to come from a mirror image of the reference beam, in relation to the hologram. An undistorted, real, three-dimensional image of the object scene appears in front of the hologram as shown in Fig. 2.12c. The conjugate image suffers from a strange depth inversion, first called "pseudoscopic" by Leith et al. [2], but it was later shown by Meier [3] that this is an incomplete description of the effect. In addition to the reversed parallax of pseudoscopic imagery, there is a reversed focus of the three-dimensional optical image. It is possible to form a hologram that will produce a real image without depth inversion by using the conjugate image of another hologram as the object, and viewing the conjugate image [4]. For plane objects such as transparencies this effect is, of course, unimportant.

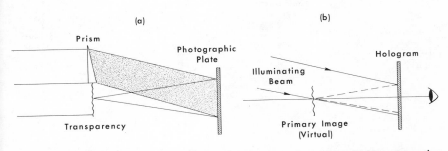

Fig. 2.13   Recording a hologram of a transparency. (*a*) Recording. (*b*) Reconstructing.

Holograms of transparencies may be made with an arrangement such as that shown in Fig. 2.13a, and the object wave reconstructed as in 2.13b. If there is no diffuser behind the transparency the hologram will appear as in Fig. 2.9. The image formed with such a hologram will be free from the characteristic "speckle" of diffuse laser illumination but suffers the disadvantage that there is almost a 1:1 correspondence between object points and position on the hologram; the hologram is essentially a shadow-gram of the object. Thus although damage to a portion of a hologram of a diffuse object has only a minor effect, damage to this hologram may result in a loss of parts of the image.

Holograms may be recorded with diverging, parallel, or converging reference beams. If care is taken to maintain the recording geometry during reconstruction, it is possible to form holograms with an arbitrary reference beam; the only requirement is that it be coherent with the object beam. A hologram formed with a nearby object is called a Fresnel hologram (Fig. 2.12). There is another class of holograms which is of primary importance. Holograms in this class are called Fraunhofer, or Fourier transform [2] holograms and are recorded as shown in Fig. 2.14a. A plane object is placed in the focal plane of a lens so that each object point gives rise to a parallel beam of light incident on the plate. The reference beam is collimated and is incident on the plate at some angle in relation to the object beam. In this way both the object and reference source are effectively at infinity. For reconstruction, a second lens is used which need not be the same focal length as the lens used in recording (Fig. 2.14b). Both the primary and conjugate images will be formed in the focal plane of this lens, magnified by the ratio $f_2:f_1$.

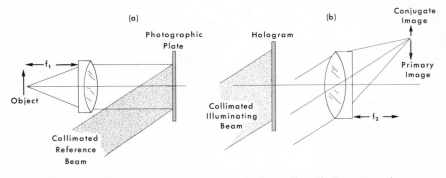

Fig. 2.14   Fourier transform hologram. (a) Recording. (b) Reconstructing.

Holograms of this type are sometimes called "Fourier transform" holograms, since if the photographic plate is placed a distance f_1 from the lens (Fig. 2.14), the light distribution at the hologram caused by the object very closely approximates the Fourier transform of the light distribution at the object (see Appendix A).

This situation can be approximated when the object and reference point are in the same plane and no lenses are used. Such a hologram has been termed a "lensless Fourier transform" hologram [5], since to a first approximation each object point gives rise to fringes of a single spatial frequency across the photographic plate (cf. Eq. 3.46). An arrangement for making such a hologram is shown in Fig. 2.15a. In Fig. 2.15b a diverging illuminating beam produces both the primary and conjugate images in a

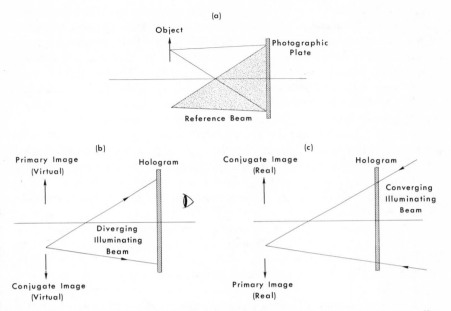

Fig. 2.15   An approximate Fourier transform hologram without lenses: (*a*) Recording. (*b*) Forming two virtual images. (*c*) Forming two real images.

plane containing the point from which the illuminating beam diverges. Both of these images are virtual. Two real images may be formed by reversing the rays in the illuminating beam as shown in Fig. 2.15*c*.

2.3   ARRANGEMENTS FOR RECORDING
VOLUME HOLOGRAMS

A volume hologram is one for which the thickness of the recording medium is of the order of or greater than the spacing of the recorded fringes. In the Leith-Upatnieks type of off-axis hologram the basic fringe spacing depends on the angle between the object and reference beams. If this angle is greater than about 7 or 8° and the recording is made with visible light, holograms made on a photographic film or plate must be considered volume holograms. In this case the fringe spacing may be of the order of $2\,\mu$, while the emulsion thickness might be $5–20\,\mu$. Therefore virtually all holograms made on photographic film should be considered volume holograms.

Volume holograms have some very useful and interesting features which are distinct from those of plane holograms. Volume holograms lend themselves very nicely to the storage of information since the whole volume of the recording medium may be utilized, as noted by van Heerden [6]. The volume hologram has been utilized by Pennington and Lin to form a three-color holographic reconstruction [7]. Denisyuk [8] introduced a new form of holography using the whole volume of the recording medium in such a way that wavefronts could be reconstructed in reflection.

These are just a few of the major useful applications of volume holograms. In all but a few situations, an accurate analysis should include considerations of the effects of the nonzero thickness of the recording medium. Plane holograms really represent a special case in the larger class of volume holograms. Except in the few special instances of truly plane holograms, all calculations assuming a two-dimensional recording medium should be considered only as a first approximation to any real experimental situation.

The main feature of volume holograms, distinct from plane holograms, is that only one image is formed for each direction of incidence of the illuminating beam. If a volume hologram of an object is formed in the same way as for a plane hologram, an aberration-free primary image (virtual) is formed in the position of the original object only if the illuminating beam is identical with the reference beam used to record the hologram. An aberration-free conjugate image (real, depth-inverted) is formed in the

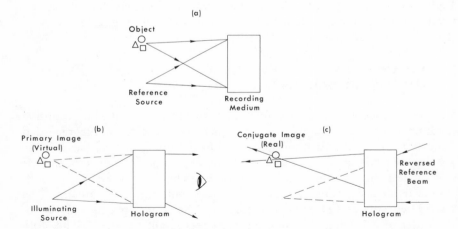

Fig. 2.16 Arrangements for recording and illuminating volume holograms. (*a*) Recording the volume hologram. (*b*) Reconstructing the primary wave. (*c*) Reconstructing the conjugate wave.

position of the object only if all of the rays in the system are reversed. These situations are shown in Fig. 2.16. In 2.16*a* a hologram is made on a thick recording medium. In 2.16*b* the primary image, which is virtual, is read out by using an exact duplicate of the reference beam for the illuminating beam. In 2.16*c* the conjugate (real) image is formed by reversing all of the rays of 2.16*a*.

Figure 2.17 shows why these arrangements are necessary. In 2.17*a*, the fringes formed by interference between light from a single object point and from the reference point are formed throughout the volume of the record-

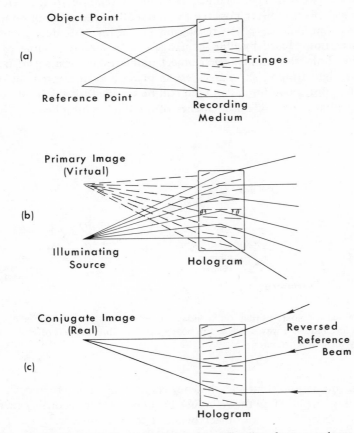

Fig. 2.17 Illustrating the Bragg condition for diffraction from a volume hologram. (*a*) Showing the formation of the fringes in the volume. (*b*) Reconstructing the primary wave. (*c*) Reconstructing the conjugate wave.

ing medium. For readout, as in 2.17*b*, the Bragg condition requires that the angle of incidence equal the angle of diffraction and at the same time both of these angles satisfy the grating equation (Chapter 4). Violation of either of these conditions will limit the amount of light diffracted and the resulting image will not appear as bright as when the Bragg condition is fulfilled. If the interference fringes are considered as reflecting planes, the Bragg condition is satisfied when the law of reflection holds and the brightest image results. Figure 2.17*c* shows how reversing the rays satisfies the Bragg condition. If the recording medium changes dimension during processing, such as by the shrinkage of photographic emulsion, it generally will not be possible to satisfy the Bragg condition during reconstruction, since the fringes will have changed position in relation to the reference beam. Since it is no longer possible to match all of the original boundary conditions simultaneously, the resolution in the image is decreased. The image distortion caused by film shrinkage can be largely eliminated with Fraunhofer holograms. Since each object point results in a set of equally spaced, straight fringes, shrinkage results in only an increased tilt of the fringes. The Bragg condition can then still be satisfied for the most part by slightly changing the angle of incidence of the illuminating beam.

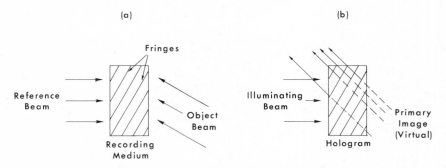

Fig. 2.18 Volume holograms that can be viewed in reflection. (*a*) Recording the hologram with object and reference waves incident in nearly opposite directions. (*b*) Reconstructing the primary wave in reflection.

Volume holograms that can be viewed in reflection are recorded as shown in Fig. 2.18*a*. Here the reference beam enters the recording medium from the side opposite the object beam. The two interfering beams are traveling in approximately opposite directions and standing waves are set up in the medium. For the arrangement shown the fringes are planes separated by approximately half the wavelength in the medium. Readout is accomplished by viewing the light reflected from these planes (Fig. 2.18*b*). Since these planes effectively form a half-wave stack interference filter, readout may be accomplished with white light, as noted by Denisyuk [8]. In this case only a small band of wavelengths from the white light source satisfies the Bragg condition at one time. The images obtained therefore appear colored and the color changes with angle of incidence. The reason for the color change with readout angle can best be understood with the aid of Fig. 2.19. Here the interference planes were recorded with wavelength λ_1 (in the medium) and so are separated by a distance $d = \lambda_1 / [2 \cos \frac{1}{2}\theta]$ (Fig. 2.19*a*). When the object wave is reconstructed by illuminating the hologram with a plane wave of white light as shown in 2.19*b*, significant reflection occurs only for a small band of wavelengths around

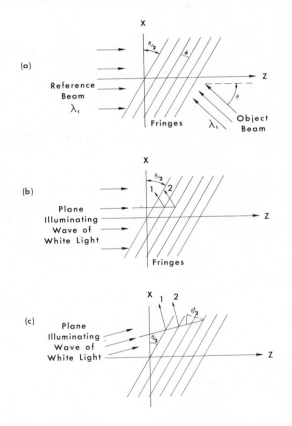

Fig. 2.19 Demonstrating the change in color of the reflected wave with a change in angle of incidence. (*a*) Recording the fringes. (*b*) Reconstructing the primary wave. (*c*) Changing the reflected color in accordance with the Bragg condition.

λ_1 because the optical path difference between a typical pair of reflected (diffracted) rays, such as (1) and (2) in the figure, is given by

$$OPD = 2d \cos (\tfrac{1}{2}\theta). \tag{2.11}$$

Since $d = \lambda_1/[2 \cos (\tfrac{1}{2}\theta)]$, rays (1) and (2) are in phase and interfere constructively, so there is strong reflection in this direction for this wavelength. If the hologram is rotated to a viewing angle of $\tfrac{1}{2}\theta'$ (Fig. 2.19c), only the wavelength satisfying the condition

$$2d \cos (\tfrac{1}{2}\theta') = \lambda_2 \tag{2.12}$$

or

$$\frac{\lambda_1}{\lambda_2} = \frac{\cos (\tfrac{1}{2}\theta)}{\cos (\tfrac{1}{2}\theta')} \tag{2.13}$$

will be strongly reflected.

Color holograms may be made using this effect. A full-color object is illuminated with red, green, and blue light, and the reference beam is composed of the same combination. Each color then forms its own set of standing waves in the medium, each with different spacing. Readout can be accomplished by illuminating with white light. Each set of Bragg planes filters out the color with which it was made and a full-color reconstruction results. A more detailed analysis of color holograms, along with other possible arrangements, will be given in Chapter 7.

The arrangements described in this chapter do not indicate all that are possible, of course. We have shown only typical examples of arrangements that fall in one or the other of the two major classifications, Fresnel holograms and Fraunhofer holograms. Generally, a Fraunhofer hologram is formed when the object is a great distance away from the recording medium. A Fresnel hologram is formed when the object is near the recording medium. A more accurate mathematical distinction between these two types will be made in Chapter 3.

REFERENCES

[1] E. N. Leith and J. Upatnieks, *J. Opt. Soc. Am.*, **52**, 1123 (1962).
[2] E. N. Leith and J. Upatnieks, *J. Opt. Soc. Am.*, **54**, 1295 (1964).
[3] R. W. Meier, *J. Opt. Soc. Am.*, **56**, 219 (1966).
[4] F. B. Rotz and A. A. Friesem, *Appl. Phys. Letters*, **8**, 146 (1966).
[5] G. W. Stroke, D. Brumm, and A. Funkhouser, *J. Opt. Soc. Am.*, **55**, 1327 (1965).
[6] P. J. van Heerden, *Appl. Opt.* **2**, 387 (1963).
[7] K. S. Pennington and L. H. Lin, *Appl. Phys. Letters*, **7**, 56 (1965).
[8] Y. N. Denisyuk, *Soviet* Phys.—*Doklady*, **7**, 543 (1962).

3 General Theory of Plane Holograms

3.0 INTRODUCTION

In this chapter the basic notation and coordinate systems to be used throughout the book will be introduced. The basic mathematics of making the hologram and reconstructing the wavefronts for each of the basic hologram types will then be discussed. In each case the fewest possible space and frequency coordinates consistent with a full understanding of the principles involved will be used. In this way the notation and mathematics will be kept as simple as possible. Extreme rigor will be avoided, but assumptions and their validity will be noted.

3.1 NOTATION

The origin of a Cartesian (x, y, z) coordinate system is centered in the recording medium (Fig. 3.1). Two-dimensional recording media (thermoplastics, thin photographic emulsions, etc.) define the x-y plane. A typical object point will be denoted by (x_o, y_o, z_o). If the reference beam is derived from a point source the coordinates of this point will be denoted by (x_R, y_R, z_R). If $z_R = \infty$, the reference beam is a plane wave and α_R will be the angle between the projection of the propagation vector $\mathbf{k}$ onto the x-z plane and the z-axis. Points in the image formed with the reconstructed wavefront will be denoted by the image coordinates (x_i, y_i, z_i) (Fig. 3.1b). During reconstruction the illuminating wave may be derived from a point source which is not identical to the reference point. We will denote the coordinates of this point by (x_c, y_c, z_c). If the reconstruction is with a plane wave, $z_c = \infty$ and we will define α_c as the angle between the projection of

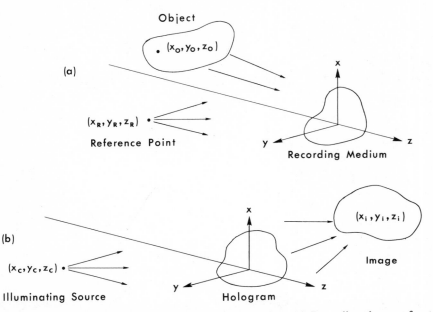

Fig. 3.1   Illustrating the coordinate system and notation. (*a*) Recording the wavefront. (*b*) Reconstructing the wavefront.

the **k** vector of the illuminating wave onto the *x-z* plane and the *z*-axis, in the same way as with the reference wave.

The amplitude distribution of the light at the object will be denoted by $F(x_o, y_o, z_o)$. This field gives rise to a disturbance $O(x, y, z)$ at the hologram. The reference wave at the hologram can be described by $R(x, y, z)$. The time dependence of these fields has been omitted, and O and R are considered to be complex amplitudes. For reconstruction the hologram is illuminated by a wave $C(x, y, z)$. The wave transmitted by the hologram in general gives rise to an image $G_p(x_i, y_i, z_i)$, or $G_c(x_i, y_i, z_i)$, denoting either the primary or conjugate image, respectively. Note that both object (F) and image (G) are also complex amplitudes. The final processed hologram will have a complex amplitude transmission $t(x, y, z)$ as a result of exposure with irradiance $I(x, y, z) = |H(x, y, z)|^2$, where

$$H(x, y, z) = O(x, y, z) + R(x, y, z) \tag{3.1}$$

is the total field at the hologram. This notation is shown in Figs. 3.2*a* and *b*.

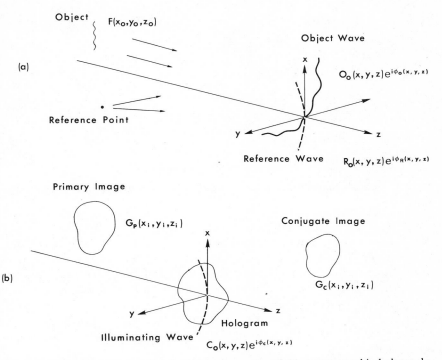

Fig. 3.2   Illustrating the notation for the various optical fields encountered in holography. (*a*) Recording the wavefront. (*b*) Reconstructing the wavefront.

3.2   ANALYSIS

3.2.1   Fresnel Holograms

Having settled on a suitable notation, we can now describe mathematically the recording of the hologram and the subsequent wavefront reconstruction. For the following discussion in this chapter, the recording medium will be considered two-dimensional. These holograms will be called *plane holograms*. Surface-deformation thermoplastics and photographic emulsions which are thin compared to the spacing of the highest frequency exposure variations are good approximations to plane holograms. A three-dimensional recording medium will be discussed in Chapter 4; these holograms will be called *volume holograms*.

Consider the arrangement shown in Fig. 3.3. The reference point at

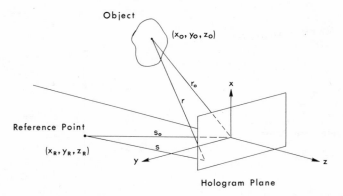

Fig. 3.3 Generalized arrangement for recording a Leith-Upatnieks off-axis, plane holo-gram.

(x_R, y_R, z_R) is the origin of a spherical reference wave. The *object* is defined by a surface S generally not including the point (x_R, y_R, z_R), no point of which lies on a line joining the reference point and any point on the record-ing medium. This is the general arrangement for a Leith-Upatnieks type of hologram [1]. A typical object point (x_o, y_o, z_o), lying on the surface S, gives rise to a spherical wave. The disturbance at the hologram plane caused by the object is the sum of the spherical waves from each point of the object:*

$$O(x, y, z) = \frac{-i}{2\lambda} \iint_{-\infty}^{\infty} F(x_o, y_o)(1 + \cos\theta)\frac{e^{ikr}}{r} \, dx_o \, dy_o. \quad (3.2)$$

The exponential expresses the change in phase of the wave traveling from (x_o, y_o, z_o) to (x, y, z); $1/r$ is the reduction in amplitude due to spreading of the spherical wave; $k = 2\pi/\lambda$ is the magnitude of the wave vector, with λ the wavelength of the light.

The factor $\frac{1}{2}(1 + \cos\theta)$ is the obliquity factor; θ is the angle between the normal to the surface at (x_o, y_o, z_o) and the line joining (x_o, y_o, z_o) with (x, y, z). The factor $-i$ appears because the Huygens' wavelets are assumed advanced in phase by $\pi/2$. The two-dimensional form of (3.2) is [see (A.1)]

$$O(x, z) = \left(\frac{-i}{4\lambda}\right)^{1/2} \int_{-\infty}^{\infty} F(x_o)(1 + \cos\theta)\frac{e^{ikr}}{r^{1/2}} \, dx_o,$$

---

* In some instances the object cannot be described on a surface, for example, a partially transmitting object occupying some volume of space. In this case the integral goes over to a volume integral if all of the point scatterers in the volume are independent or if the scattering is weak. (See [2] and [3] for a more complete treatment of Huygens' principle.)

and this is the form which we will use throughout most of this book in order to simplify many of the equations. For the purpose of classifying hologram types, however, we return to (3.2) and write the distance r as

$$r = [(x_o - x)^2 + (y_o - y)^2 + (z_o - z)^2]^{1/2}. \tag{3.3}$$

The reference wave at the hologram plane is given by

$$R(x, y, z) = R_o \cdot \frac{e^{iks}}{s} \tag{3.4}$$

with

$$s = [(x_R - x)^2 + (y_R - y)^2 + (z_R - z)^2]^{1/2}. \tag{3.5}$$

For plane holograms, we are interested only in the plane $z = 0$, therefore

$$r = [(x_o - x)^2 + (y_o - y)^2 + z_o^2]^{1/2}. \tag{3.6}$$

In many cases in holography, $z_o^2 \sim x_o^2 + y_o^2$, and we may not use the binomial expansion for (3.6). In this case, the integral (3.2) is extremely difficult and analysis is usually not performed explicitly. When $z_o^2 > x_o^2$, y_o^2, we may expand Eq. 3.6 as

$$r \cong z_o + \frac{x_o^2 + y_o^2}{2z_o} + \frac{x^2 + y^2}{2z_o} - \frac{xx_o}{z_o} - \frac{yy_o}{z_o} - \frac{(x_o - x)^4}{8z_o^3} - \frac{(y_o - y)^4}{8z_o^3} + \cdots. \tag{3.7}$$

When we can neglect quadratic and higher order terms in x_o and y_o, we speak of *Fraunhofer holograms* (for plane reference waves); when the quadratic terms cannot be neglected, we speak of *Fresnel holograms*. We will consider the latter first.

If we write $O(x, y) = O_o(x, y)e^{i\varphi_o(x,y)}$ and $R(x, y) = R_o(x, y)e^{i\varphi_R(x,y)}$, where O_o and R_o are real and φ_o and φ_R describe the spatial phase variation of the object and reference beams, respectively, then Eq. 3.1 becomes

$$H(x, y) = O_o e^{i\varphi_o} + R_o e^{i\varphi_R}. \tag{3.8}$$

From (2.2), this field produces an exposure

$$E(x, y) = |H(x, y)|^2 \tag{3.9}$$

which gives rise to a final amplitude transmittance

$$t(x, y) = \beta |H(x, y)|^2$$
$$= \beta[O_o^2 + R_o^2 + O_o R_o e^{i(\varphi_o - \varphi_R)} + O_o R_o e^{-i(\varphi_o - \varphi_R)}]. \tag{3.10}$$

If we illuminate the hologram with a wave

$$C(x, y) = C_o(x, y)e^{i\varphi_c(x,y)}, \tag{3.11}$$

the transmitted wave at the hologram will be

$$\psi(x, y) = C(x, y) \cdot t(x, y)$$
$$= \beta[C_o O_o^2 e^{i\varphi_c} + C_o R_o^2 e^{i\varphi_c} + C_o O_o R_o e^{i(\varphi_c + \varphi_o - \varphi_R)}$$
$$+ C_o O_o R_o e^{i(\varphi_c - \varphi_o + \varphi_R)}]. \qquad (3.12)$$

This is the basic equation of holography. The way in which the phases of the various terms are expressed determines the type of hologram, either Fresnel or Fraunhofer.

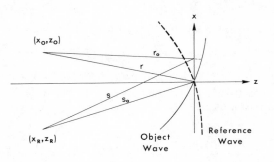

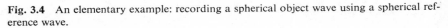

Fig. 3.4 An elementary example: recording a spherical object wave using a spherical reference wave.

To see more clearly how the wavefront reconstruction process works, let us analyze in detail a specific example. A suitable simple arrangement is shown in Fig. 3.4. Here we are considering a single point object and a single point reference source. Both sources emit spherical waves, therefore

$$O_o(x, y)e^{i\varphi_o(x,y)} = B \frac{e^{ikr}}{r} \qquad (3.13)$$

and

$$R_o(x, y)e^{i\varphi_R(x,y)} = A \frac{e^{iks}}{s}; \qquad (3.14)$$

A and B are the initial amplitude of the reference and object waves, respectively. This result can be obtained with the use of (3.2); e.g., see [3], p. 369. Assume that A and B are essentially constant over the area of the hologram. The distances r_o, r, s_o, and s are given by

$$r = [(x - x_o)^2 + (y - y_o)^2 + z_o^2]^{\frac{1}{2}} \qquad (3.15)$$
$$r_o = (x_o^2 + y_o^2 + z_o^2)^{\frac{1}{2}} \qquad (3.16)$$

and

$$s = [(x - x_R)^2 + (y - y_R)^2 + z_R^2]^{1/2} \tag{3.17}$$

$$s_o = (x_R^2 + y_R^2 + z_R^2)^{1/2}. \tag{3.18}$$

From (3.13) and (3.15) we have

$$\varphi_o(x, y) = k(r - r_o)$$
$$= \frac{2\pi}{\lambda} \left\{ [(x - x_o)^2 + (y - y_o)^2 + z_o^2]^{1/2} - (x_o^2 + y_o^2 + z_o^2)^{1/2} \right\} \tag{3.19}$$

which is the phase of the object wave in the hologram plane relative to the origin. Expansion of the square roots yields

$$\varphi_o(x, y) = \frac{2\pi}{\lambda} \left[\frac{1}{2z_o} (x^2 + y^2 - 2xx_o - 2yy_o) + \cdots \right]. \tag{3.20}$$

The terms that have been omitted in (3.20) contain $(1/z_o)^3$, $(1/z_o)^5$, and so on. For our present purposes we are interested only in the first-order terms; the higher order terms in some cases are not negligible, however, and they give rise to aberrations of the reconstructed wave. These higher order terms will be discussed in Chapter 5.

A similar expression is obtained for the phase of the reference wave

$$\varphi_R(x, y) = \frac{2\pi}{\lambda} \left[\frac{1}{2z_R} (x^2 + y^2 - 2xx_R - 2yy_R) + \cdots \right]. \tag{3.21}$$

Similarly, for the illuminating wave,

$$\varphi_c(x, y) = \frac{2\pi}{\lambda} \left[\frac{1}{2z_c} (x^2 + y^2 - 2xx_c - 2yy_c) + \cdots \right]. \tag{3.22}$$

In the third term of Eq. 3.12, these phases add in the combination

$$\varphi_o - \varphi_R + \varphi_c = \frac{\pi}{\lambda} \left[(x^2 + y^2) \left(\frac{1}{z_o} - \frac{1}{z_R} + \frac{1}{z_c} \right) \right.$$
$$\left. - 2x \left(\frac{x_o}{z_o} - \frac{x_R}{z_R} + \frac{x_c}{z_c} \right) - 2y \left(\frac{y_o}{z_o} - \frac{y_R}{z_R} + \frac{y_c}{z_c} \right) \right]. \tag{3.23}$$

Following Meier [4], we now consider this as representing the first-order term of the reconstructed spherical wave,

$$\Phi_p^{(1)} = \frac{\pi}{\lambda} \left(\frac{x^2 + y^2 - 2xX_p - 2yY_p}{Z_p} \right) \tag{3.24}$$

with Z_p the radius and X_p, Y_p the center coordinates of the spherical wave.

The subscript p indicates that we are referring to the primary wave. From (3.23) and (3.24),

$$\mathbf{Z}_p = \frac{z_o z_R z_c}{z_R z_c - z_o z_c + z_o z_R}$$

$$\mathbf{X}_p = \frac{x_o z_R z_c - x_R z_o z_c + x_c z_o z_R}{z_R z_c - z_o z_c + z_o z_R} \qquad (3.25)$$

$$\mathbf{Y}_p = \frac{y_o z_R z_c - y_{Ro} z_o z_c + y_c z_o z_R}{z_R z_c - z_o z_c + z_o z_R}$$

If the illuminating wave is identical with the reference wave, then $z_c = z_R$, $x_c = x_R$, $y_c = y_R$ and

$$\mathbf{Z}_p = z_o; \qquad \mathbf{X}_p = x_o; \qquad \mathbf{Y}_p = y_o \qquad (3.26)$$

so that the reconstructed wave exactly corresponds to the original object wave.

Performing the same set of operations for the fourth term of Eq. 3.12 yields

$$\mathbf{Z}_c = \frac{z_o z_R z_c}{z_o z_R - z_c z_R + z_c z_o}$$

$$\mathbf{X}_c = \frac{x_c z_o z_R - x_o z_c z_R + x_R z_c z_o}{z_o z_R - z_c z_R + z_c z_o} \qquad (3.27)$$

$$\mathbf{Y}_c = \frac{y_c z_o z_R - y_o z_c z_R + y_R z_c z_o}{z_o z_R - z_c z_R + z_c z_o}$$

where the subscript c means that we are referring to the conjugate wave. For the illuminating wave identical with the reference wave, $z_R = z_c$, $x_R = x_c$, $y_R = y_c$, then

$$\mathbf{Z}_c = -z_o$$

$$\mathbf{X}_c = \frac{2x_R z_o - x_o z_R}{2z_o - z_R} \qquad (3.28)$$

$$\mathbf{Y}_c = \frac{2y_R z_o - y_o z_R}{2z_o - z_R}$$

so that this wave yields an image on the opposite side of the hologram plane from the object, and displaced from the axis by an amount which depends on the original object and reference point coordinates.

Whether the images are real or virtual depends on the sign of $\mathbf{Z}_p$ and $\mathbf{Z}_c$. Since z_o is always negative (because of the way in which we have set up the coordinate system) a negative $\mathbf{Z}_p$ or $\mathbf{Z}_c$ means a virtual image. Since z_R and z_c are arbitrary, both images may be real or virtual.

When the illuminating wave is not identical to the reference wave, the first-order image positions are given by (3.25) and (3.27). We shall see later that this gives rise to magnification along with aberrations. Magnification and wavefront aberrations also occur when the wavelength of illuminating beam is not the same as the wavelength of the reference beam.

One very common arrangement used in holography involves a plane reference wave. In this case $z_R = \infty$ and we denote the direction of the wave by α_R, the angle between the propagation vector $\mathbf{k}$ and the z-axis in the x-z plane. Analogously, we reconstruct with a plane wave ($z_c = \infty$) at an angle α_c (Fig. 3.5). In this situation we have

$$\varphi_R = \frac{2\pi}{\lambda} x \sin \alpha_R$$

$$\varphi_c = \frac{2\pi}{\lambda} x \sin \alpha_c \qquad (3.29)$$

so that

$$\varphi_o - \varphi_R + \varphi_c = \frac{\pi}{\lambda} \left[\frac{x^2 + y^2}{z_o} - 2x \left(\frac{x_o}{z_o} - \sin \alpha_R + \sin \alpha_c \right) - 2y \frac{y_o}{z_o} \right] \quad (3.30)$$

and

$$\mathbf{Z}_p = z_o$$

$$\mathbf{X}_p = x_o - z_o \sin \alpha_R + z_o \sin \alpha_c \qquad (3.31)$$

$$\mathbf{Y}_p = y_o.$$

For the illuminating beam identical with the reference beam ($\sin \alpha_R = \sin \alpha_c$), one of the reconstructed waves exactly corresponds to the original object wave. For the conjugate image we have

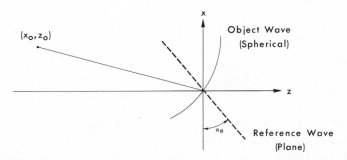

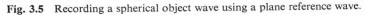
Fig. 3.5 Recording a spherical object wave using a plane reference wave.

$$\varphi_R - \varphi_o + \varphi_c$$

$$= \frac{2\pi}{\lambda} \left[x \sin \alpha_R - \frac{1}{2z_o}(x^2 + y^2 - 2xx_o - 2yy_o) + x \sin \alpha_c \right] \quad (3.32)$$

and

$$\mathbf{Z}_c = -z_o$$

$$\mathbf{X}_c = x_o + z_o \sin \alpha_R + z_o \sin \alpha_c \qquad (3.33)$$

$$\mathbf{Y}_c = y_o;$$

therefore both images are displaced from the axis and from each other by amounts which depend on the offset angle of reference and illuminating beams.

3.2.2   Fraunhofer Holograms

The designation *Fresnel hologram* has been applied where the terms of order two or greater in x_o and y_o may be neglected in the expansion (3.7). Generally, a Fresnel hologram is formed when the object is near the hologram plane. When the object size is small compared to its distance z_o from the hologram plane, we need consider only the first-order terms in x_o and y_o in Eq. 3.7. In this case, we speak of a Fraunhofer hologram if the reference wave is plane. The obliquity factor of Eq. 3.2 is nearly unity and

$$r \sim z_o + \frac{x^2 + y^2}{2z_o} - \frac{xx_o}{z_o} - \frac{yy_o}{z_o}. \qquad (3.34)$$

The denominator r in the integral of (3.2) may be written as z_o and taken outside the integral; therefore (3.2) becomes

$$O(x, y) = \frac{-i}{2\lambda z_o} e^{ikz_o} \exp\left[ik \frac{(x^2 + y^2)}{2z_o} \right]$$

$$\times \iint_{-\infty}^{\infty} F(x_o, y_o) \exp\left[-\frac{ik}{z_o}(xx_o + yy_o) \right] dx_o\, dy_o \quad (3.35)$$

for a *plane* hologram. Strictly speaking, we can neglect the second and higher order terms in x and y only in the limiting case $z_o \to \infty$. Equation 3.35, however, is valid for the situation where a large, well-corrected lens is placed a focal length away from the object and the x, y plane is in the rear focal plane of the lens. To understand why this is true, consider the two situations shown in Fig. 3.6. In Fig. 3.6a we see that the disturbance in the x, y plane, a great distance from the object, can be considered as arising from the superposition of plane waves originating from each point of the object and traveling in the direction defined by the angle α_o. If a well-corrected lens is placed a focal length away from the object (Fig. 3.6b), however, all of the light from the object traveling at an angle α_o will come to a focus at a point x in the focal plane of the lens. Since the optical path from the wavefront of the wave traveling at an angle α_o to x is the same for all rays, we obtain essentially the same interference effects predicted by (3.35). The reason for placing the lens a distance f (one focal length) from the object is discussed in Appendix A.

Hence we can write Eq. 3.35 as

$$O(x, y) = C \iint_{-\infty}^{\infty} F(x_o, y_o) \exp\left[-i\frac{k}{f}(xx_o + yy_o) \right] dx_o \, dy_o \quad (3.36)$$

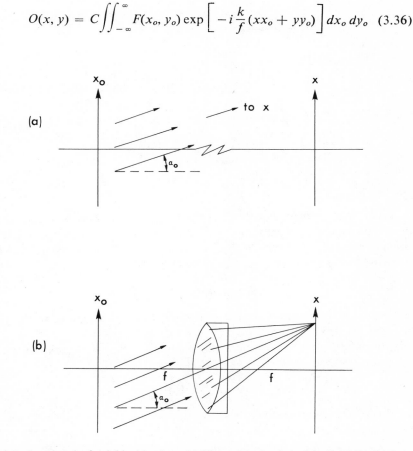

Fig. 3.6 Imaging the far field with a lens. (a) Illustrating the desired far field distribution. (b) Showing how the desired field distribution forms in the focal plane of a lens.

where $C = -(i/2z_o\lambda)e^{ikz_o}$. By suitably defining the object function $F(x_o, y_o)$ so that it goes to zero beyond the object, the limits on the integral can be extended to $\pm\infty$. In this case, we see that, apart from a constant, the amplitude distribution of the hologram plane from the object is very nearly the two-dimensional Fourier transform of the object distribution (see Appendix A). This same result is obtained, of course, when z_o is very large compared with $(x^2 + y^2)^{1/2}$. Holograms of this sort are often called "Fourier transform" holograms, when the reference wave is plane.

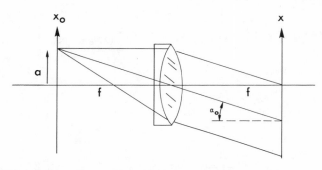

Fig. 3.7   Illustrating the formation of the object wave $O(x)$ in a Fourier transform hologram system. A single object point is located at $x_o = a$.

This type of hologram has some very interesting properties. Suppose the object consists of a single point at $x_o = a$ (considering only the two-dimensional example as in Fig. 3.7). Then $F(x_o, y_o) = \delta(x_o - a)$, where $\delta(x)$ denotes the Dirac delta function (see ref. [3], p. 752), and

$$O(x) = \text{const} \times e^{-i(k/f)ax} \qquad (3.37)$$

This represents a plane wave incident on the hologram plane at an angle $\alpha_o = \sin^{-1}(a/f)$.

If the reference wave is now a plane wave striking the hologram plane at an angle α_R, we have

$$\delta(x) \equiv \varphi_R(x) - \varphi_o(x) = kx(\sin\alpha_R - \sin\alpha_o) \qquad (3.38)$$

for the phase difference between the two waves. There will be a bright fringe for $\delta = 2m\pi$ where m is an integer, or for

$$x = \frac{m(2\pi/k)}{\sin\alpha_R - \sin\alpha_o} \qquad (3.39)$$

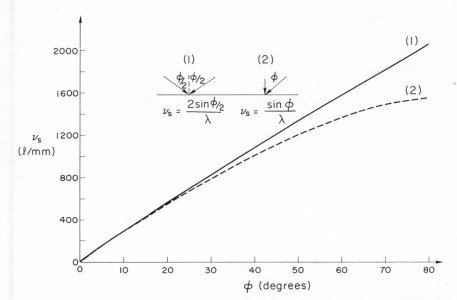

Fig. 3.8 The spatial frequency of the interference fringes ν_s as a function of the total angular separation φ between the object and reference beams for two common configurations.

where $k = 2\pi/\lambda$. Thus a single object point in a Fourier transform system forms a set of straight fringes of spacing

$$\Delta x_f = \Delta m \frac{\lambda}{\sin \alpha_R - \sin \alpha_o} = \frac{\lambda}{\sin \alpha_R - \sin \alpha_o}.$$

The inverse of this fringe spacing, $1/\Delta x_f$, is the basic spatial frequency recorded on the hologram. Figure 3.8 shows how the spatial frequency ν_s varies as a function of the total angle φ between the two beams for two common situations. In (1), the normal to the recording plane bisects the angle formed by the two beams so that $\alpha_o = -\alpha_R$; in (2), one of the beams is incident normally, $\alpha_o = 0$. The wavelength used for the computation is the He-Ne laser wavelength, $.63 \mu$. If the hologram is a Fresnel hologram, there are, of course, many values of α_o—in this case the curves of Fig. 3.8 are still useful if we call α_o the angle subtended at the hologram plane by the center of the object—ν_s will then be the average spatial frequency of the hologram.

Returning to the problem at hand, we have, as usual, the total field at the hologram given by (in one dimension)

$$H(x) = O_o(x)e^{i\varphi_o(x)} + R_o(x)e^{i\varphi_R(x)} \tag{3.40}$$

which can be written

$$H(x) = O_o e^{-ikx\sin\alpha_o} + R_o e^{-ikx\sin\alpha_R}, \tag{3.41}$$

so that the exposure is [by (3.9)]

$$E(x) = [O_o^2 + R_o^2 + O_o R_o e^{-ikx(\sin\alpha_o - \sin\alpha_R)} + O_o R_o e^{-ikx(\sin\alpha_R - \sin\alpha_o)}]. \tag{3.42}$$

We will assume the amplitude transmittance to be given as in Eq. 2.2: $t(x) = \beta E(x)$. If we illuminate the hologram with a plane wave

$$C(x) = C_o(x)e^{i\varphi_c(x)} = C_o e^{-ikx\sin\alpha_c} \tag{3.43}$$

then the transmitted wave is of the form

$$\begin{aligned}
\psi(x) = \beta[&C_o O_o^2 e^{-ikx\sin\alpha_c} + C_o R_o^2 e^{-ikx\sin\alpha_c} \\
&+ C_o O_o R_o e^{-ikx(\sin\alpha_o - \sin\alpha_R + \sin\alpha_c)} \\
&+ C_o O_o R_o e^{-ikx(\sin\alpha_R - \sin\alpha_o + \sin\alpha_c)}]. \tag{3.44}
\end{aligned}$$

Each of these terms represents a plane wave, since the phase advances linearly with x. The two waves represented by the first two terms are traveling in the same direction as the illuminating wave. This is the zero-order wave from the gratinglike pattern on the hologram. The third and fourth terms are equal amplitude plane waves traveling in two different directions which are symmetrical about the zero order. These represent the two first orders of the grating and they both image the object point at infinity. If $\alpha_c = \alpha_R$, then we see that the third term represents the reconstructed object wave. If the recording is linear, as has been assumed, there are no higher diffracted orders since the grating transmittance is sinusoidal. This can be seen from (3.42) which can be written

$$E(x) = \{O_o^2 + R_o^2 + 2O_o R_o \cos[kx(\sin\alpha_o - \sin\alpha_R)]\}. \tag{3.45}$$

Figures 3.9a and b show schematically the recording and reconstructing of a single-point Fourier transform hologram.

The extension to more than a single object point is simple if we extend the diffraction grating analogy. Each object point yields a plane wave at the hologram plane which interferes with a plane reference wave, producing a multiplicity of coherently superposed sinusoidal gratings. Each grating then forms two diffracted first orders which form the two images at infinity of each object point. The modulation, or fringe visibility, of each of the component gratings decreases as the number of object points increases, but as we will see in Chapter 8 when discussing information storage capabilities, the allowable object size (or number of object points) becomes

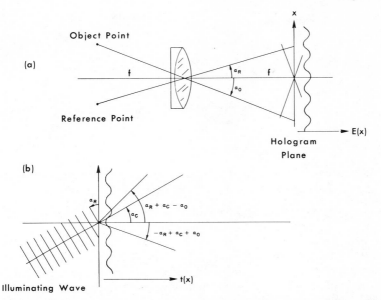

Fig. 3.9 The Fourier transform hologram. (*a*) The wavefront from a single object point is recorded as a grating which varies sinusoidally in transmittance. (*b*) The object wave is reconstructed as one of the first-order diffractions from the hologram.

quite large before the modulation decreases to a point where no hologram can be recorded.

There is one other special recording arrangement for producing Fourier transform holograms which deserves mention here. If the reference beam is derived from a point source that is in the same plane as the object, and no lenses are used, the sphericities of the two spherical waves at the hologram plane (one from an object point, the other from the reference point) tend to cancel. Thus an approximately constant-spatial-frequency grating is produced for each object point and we have a "lensless Fourier transform hologram" [5]. To see how this comes about, consider the arrangement of Fig. 3.10. For an object point at x_o and the reference point at x_R, the angle between the two rays at the hologram is $\beta(x)$, where

$$\beta(x) = \tan^{-1}\left(\frac{x_o - x}{z_o}\right) + \tan^{-1}\left(\frac{x - x_R}{z_o}\right)$$

$$= \frac{x_o - x_R}{z_o} - \frac{1}{3z_o^3}[x_o^3 - x_R^3 - 3x(x_o^2 - x_R^2) + 3x^2(x_o - x_R)] + \cdots$$

$$(3.46)$$

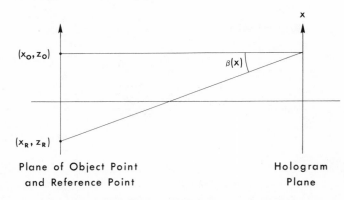

Fig. 3.10   Illustrating the lensless Fourier transform hologram.

which does not vary strongly with x for $x/z_o \ll 1$. Since the angle between the two interfering beams is approximately constant across the hologram, so is the spacing of the fringes: hence we have an approximate Fourier transform hologram, although not in the sense that the field distribution at the hologram caused by the object is the Fourier transform of the object distribution.

3.3   PHASE HOLOGRAMS

3.3.1   Introduction

Thus far we have considered only absorption holograms, for which the light of the illuminating beam is absorbed in correspondence with the recording exposure. We have assumed that the transmission function $t(x)$ was a real valued function. We will now show that holograms can also be formed on recording media for which $t(x)$ is a complex function, that is, the hologram alters the *phase* of the illuminating wave in correspondence with the recording exposure. Such a hologram is termed a "phase hologram" and has some interesting and important properties.

In order to demonstrate most clearly the principles of phase holograms, we will consider only the case of pure phase modulation wherein $t(x)$ is pure imaginary. Holograms can be produced with $t(x)$ having both real and imaginary parts—some absorption and some phase modulation (indeed most of the holograms produced are probably of this type since it is difficult to eliminate phase modulation completely).

Pure phase holograms can be made by many methods. The earliest reference to phase holograms describes their preparation by contact printing of Gabor holograms, employing the Carbro process to produce relief images in transparent gelatin on glass substrates [6]. Phase holograms may be produced by contact printing (or by direct recording with blue or green laser light) onto a resist material. They may be produced by using a thermoplastic material as the recording medium [7]. It is also possible to make phase holograms with conventional silver halide photographic emulsions—either by utilizing the relief image or index change, or both. Relief images are an imagewise emulsion thickness variation caused by the differential tanning action of the developer [8]. These thickness variations result in the desired spatial phase modulation of the transmitted light. To obtain a pure phase modulation using this technique, one can bleach out the exposed silver grains so that no density variation remains [9], or one can metallize the emulsion surface and view the hologram image in reflection [10]. Phase holograms may be produced by utilizing a change in the index of refraction of the recording medium. In the photoresist materials a change in index results from a photopolymerization of the resist material upon exposure. There are now no exposurewise thickness variations, but the signal is recorded in the form of index variations that impose the desired phase modulation onto the illuminating beam. A bleached photographic hologram also shows evidence of index change in the exposed areas.

These are just some of the methods for producing phase holograms which are known at this time. There will undoubtedly be more methods. The main interest in phase holograms stems from their very high *diffraction efficiency*—the ratio of incident to usable diffracted flux. Burckhardt has shown that the efficiency of a thick dielectric grating can be as high as 100% [11]. For a thin phase grating, the efficiency can be as high as 34% [12]. Compare these numbers with a little over 6% for the usual absorption type of gratings (or hologram).

3.3.2 Analysis

For simplicity, we will restrict our attention to pure phase holograms—those for which only the phase and not the amplitude of the illuminating wave is affected.

To begin, we suppose that the recording medium is such that an exposure $E(x)$ results in either a change in index $n(x)$ or a change in thickness $h(x)$ of the medium. If the exposed hologram is illuminated with a wave $C(x)$, then the transmitted wave will have a phase variation imposed on it of the form $knh(x)$ or $khn(x)$ for the relief or index types, respectively. This imposed phase modulation yields various diffracted orders, the $\pm$1st orders producing the desired reconstructions. For thick holograms, one can arrange to produce only one diffracted order, *and* one can eliminate the zero order by a suitable adjustment of the phase shift.

Next we expose the recording medium to an object wave described by

$$O(x) = O_o e^{i\varphi_o(x)} \tag{3.47}$$

and a plane reference wave

$$R(x) = R_o e^{i\beta x} \tag{3.48}$$

where $\beta = k \sin \alpha_R$, as shown in Fig. 3.11. Let us further assume that the final phase modulation to be imposed on the illuminating wave $C(x)$ is

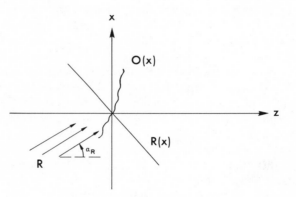

Fig. 3.11   Recording a hologram.

simply equal to the exposure $E(x)$. Call this phase modulation $\varphi(x)$ so that we can write

$$\varphi(x) = E(x), \tag{3.49}$$

where $E(x)$ can be written as (ignoring the proportionality constant)

$$
\begin{aligned}
E(x) &= |O(x) + R(x)|^2 \\
&= O_o^2 + R_o^2 + O_o R_o e^{-i(\varphi_o - \beta x)} + O_o R_o e^{i(\varphi_o - \beta x)} \\
&= O_o^2 + R_o^2 + 2 O_o R_o \cos(\varphi_o - \beta x). \tag{3.50}
\end{aligned}
$$

The complex transmittance function then becomes

$$
\begin{aligned}
t(x) &= t_o e^{i\varphi(x)} = t_o e^{iE(x)} \\
&= t_o e^{iO_o^2} e^{iR_o^2} e^{2iO_o R_o [\cos(\varphi_o - \beta x)]} \tag{3.51}
\end{aligned}
$$

For simplification of notation, we write

$$K \equiv t_o e^{iO_o^2} e^{iR_o^2}$$

$$a \equiv 2O_o R_o \qquad (3.52)$$

$$\theta \equiv \varphi_o - \beta x$$

so that

$$t(x) = K e^{ia\cos\theta} \qquad (3.53)$$

Using the Bessel function expansions

$$\cos\left(a\cos\theta\right) = J_o(a) + 2\sum_{n=1}^{\infty}(-1)^n J_{2n}(a)\cos\left(2n\theta\right)$$

$$\sin\left(a\cos\theta\right) = 2\sum_{n=0}^{\infty}(-1)^{n+2} J_{2n+1}(a)\cos\left[(2n+1)\theta\right] \qquad (3.54)$$

and the relation

$$e^{ix} = \cos x + i\sin x \qquad (3.55)$$

we can write

$$t(x) = K\left\{J_o(a) + 2\sum_{n=1}^{\infty}(-1)^n J_{2n}(a)\cos\left(2n\theta\right)\right.$$

$$\left. + 2i\sum_{n=0}^{\infty}(-1)^{n+2} J_{2n+1}(a)\cos\left[(2n+1)\theta\right]\right\}. \qquad (3.56)$$

The $J_n(a)$ are Bessel functions of the first kind. Each $J_n(a)$ is the amplitude of the nth diffracted order. Note that in general all orders are present, unlike the case of an absorption hologram where the sinusoidal amplitude modulation led to only the ±1st orders. The term of (3.56) leading to the images of interest is

$$K[2iJ_1(a)\cos\theta], \qquad (3.57)$$

which can be written as

$$2iKJ_1(a)\left(\frac{e^{i\theta} + e^{-i\theta}}{2}\right) = iKJ_1(a)[e^{i(\varphi_o - \beta x)} + e^{-i(\varphi_o - \beta x)}]. \qquad (3.58)$$

These two terms represent the primary and conjugate image waves. The transmission term leading to the primary image is

$$t_p(x) = t_o \exp\left[i\left(O_o^2 + R_o^2 + \frac{\pi}{2}\right)\right]J_1(2O_o R_o)e^{i(\varphi_o - \beta x)}. \qquad (3.59)$$

If the hologram is illuminated with a wave $C(x) = R(x)$ (Fig. 3.12) we obtain the primary image wave

$$\psi_p(x) = C(x)t_p(x) = R_o e^{i\beta x}t_p(x)$$
$$= t_o e^{i(O_o{}^2+R_o{}^2+\pi/2)}R_o J_1(2O_o R_o)e^{i\varphi_o}. \tag{3.60}$$

Since

$$J_1(x) = \tfrac{1}{2}x - \frac{(\tfrac{1}{2}x)^3}{1^2\cdot 2} + \frac{(\tfrac{1}{2}x)^5}{1^2\cdot 2^2\cdot 3} - \cdots$$

$$\approx \tfrac{1}{2}x, \qquad \text{for small } x, \tag{3.61}$$

we have, if $2O_o R_o$ is small,

$$\psi_p(x) \cong \frac{t_o}{2} e^{i(O_o{}^2+R_o{}^2+\pi/2)}R_o{}^2 O_o e^{i\varphi_o}, \tag{3.62}$$

which is seen to be, aside from the unimportant phase and amplitude factors, just the original object wave. If the product $2O_o R_o$ is large, some amplitude distortion will be present. From (3.49) and (3.50), we see that $2O_o R_o$ is the amplitude of the phase modulation

$$\varphi(x) = O_o{}^2 + R_o{}^2 + 2O_o R_o \cos (\varphi_o - \beta x). \tag{3.63}$$

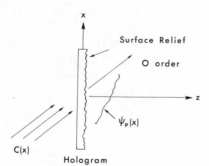

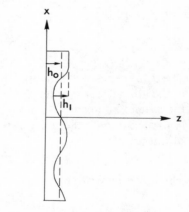

Fig. 3.12 Diffraction from a phase hologram.

Fig. 3.13 Relief image parameters.

Hence if the phase hologram is of the relief image type we might have, for example,

$$\varphi(x) = knh(x) = O_o{}^2 + R_o{}^2 + 2O_o R_o \cos (\varphi_o - \beta x) \tag{3.64}$$

which can be written in the form

$$\varphi(x) = kn\left[h_o + \frac{h_1}{2}\cos(\varphi_o - \beta x)\right] \qquad (3.65)$$

where h_1 is the peak-to-peak variation of the hologram thickness and the average thickness is h_o, as shown in Fig. 3.13. A distortion-free primary (or conjugate) image may be formed with a phase hologram if the phase modulation is small; in the foregoing example, h_1 should be kept small.

REFERENCES

[1] E. N. Leith and J. Upatnieks, *J. Opt. Soc. Am.*, **52,** 1123 (1962).

[2] B. B. Baker and E. T. Copson, *The Mathematical Theory of Huygen's Principle*, Oxford University Press, London, 1950.

[3] M. Born and E. Wolf, *Principles of Optics*, Pergamon Press Ltd., Oxford, 1964.

[4] R. W. Meier, *J. Opt. Soc. Am.*, **55,** 987 (1965).

[5] G. W. Stroke, D. Brumm, and O. Funkhouser, *J. Opt. Soc. Am.*, **55,** 1327 (1965).

[6] G. L. Rogers, *Proc. Roy. Soc.*, Edinburgh, **A63,** 193 (1952).

[7] J. C. Urbach and R. W. Meier, *Appl. Opt.*, **5,** 666 (1966).

[8] H. M. Smith, *J. Opt. Soc. Am.*, **58,** 533 (1968).

[9] W. T. Cathey, Jr., *J. Opt. Soc. Am.*, **55,** 457 (1965).

[10] A. K Rigler, *J. Opt. Soc. Am.*, **55,** 1693 (1965).

[11] C. B. Burckhardt, *J. Opt. Soc. Am.*, **57,** 601 (1967).

[12] H. Kogelnik, *Microwaves*, 68, November 1967.

4 General Theory of Volume Holograms

4.0 INTRODUCTION

Volume holograms, as defined in Chapter 3, are those for which the thickness of the recording medium is not negligible compared to the fringe spacing. In the Leith-Upatnieks off-axis type of hologram, with the images well separated in space, the information recorded on the hologram is contained in a spatial frequency band centered on the spatial carrier frequency determined by the offset angle (see Fig. 3.8). It is the fringe spacing associated with this carrier frequency that determines whether a hologram is a plane hologram or a volume hologram. If the center of the object is angularly separated from the reference point by an angle φ, the fringe spacing of the carrier frequency is of the order of

$$d \sim \frac{\lambda}{\sin \varphi} \tag{4.1}$$

where λ is the wavelength of the recording light. When d is of the order of, or smaller than, the thickness of the recording medium, the hologram must be considered a volume hologram. By far the most common material used for recording is photographic film with emulsion thicknesses ranging from a few to 20 μ and more. Therefore most holograms recorded on film should be considered volume holograms. In the gray area of low carrier frequency holograms on thin emulsions, most of the equations for plane holograms will be valid enough, but a careful analysis of any experiment should include consideration of the nonzero emulsion thickness. Because of the greatly increased difficulty of solving the three-dimensional problem, most analyses to date have considered only the two-dimensional problem. For many experiments these can be considered reasonable first approximations. For surface deformation thermoplastics and relief image holograms the two-dimensional solutions are accurate.

4.1   FRINGES IN THREE DIMENSIONS

To begin our analysis of volume holograms, we note that the fringes formed by the interference of the object beam and the reference beam are located throughout the depth of the recording medium. The direction and spacing of the fringes depend on the location and angular separation of the object and reference point. In general, for a single reference point each object point will give rise to fringes in the recording medium whose spacing and direction vary with position. For object and reference points both at infinity, but angularly separated, the fringe spacing and direction will be constant. Since an arbitrary object wave may be decomposed into a spectrum of plane waves, we will consider only the simple problem of finding the spacing and orientation of the fringes caused by the interference of two plane waves incident on a three-dimensional recording medium.

Consider the situation shown in Fig. 4.1. Here we are assuming that the **k** vectors of the two interfering beams lie in the x-z plane. The coordinate system is centered in the recording medium and all directions refer to directions in the medium. The associated directions outside the medium may be found by using Snell's law. Let the two interfering waves be denoted by u_o and u_R. The directions of propagation of these two waves in the medium make angles θ_o and θ_R with the z-axis, respectively; the positive direction is as shown. Let $k' = 2\pi/\lambda'$ (with λ' the wavelength in the medium) be the common wave number of the two waves, and $\hat{k}_o$ and $\hat{k}_R$ be unit vectors in the direction of propagation so that

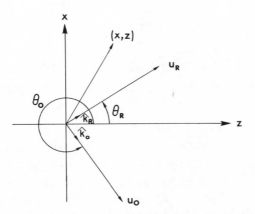

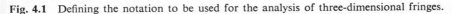

Fig. 4.1 Defining the notation to be used for the analysis of three-dimensional fringes.

$$\mathbf{k}_o' \equiv \hat{\mathbf{k}}_o k'$$
$$\mathbf{k}_R' \equiv \hat{\mathbf{k}}_R k' \qquad (4.2)$$

are the wave vectors of the two waves. If $\mathbf{r}$ is the vector position of a point (x, z) in the medium, then we may write

$$u_o = a_o e^{i\mathbf{k}'_o \cdot \mathbf{r}} = a_o e^{i\varphi_o(x, z)}$$
$$u_R = a_R e^{i\mathbf{k}'_R \cdot \mathbf{r}} = a_R e^{i\varphi_R(x, z)}. \qquad (4.3)$$

If we define $\hat{\mathbf{x}}$ and $\hat{\mathbf{z}}$ as unit vectors along the x and z axes, respectively, then

$$\mathbf{r} = \hat{\mathbf{x}} x + \hat{\mathbf{z}} z \qquad (4.4)$$

and therefore

$$\hat{\mathbf{k}}_o = \hat{\mathbf{z}} \cos \theta_o + \hat{\mathbf{x}} \sin \theta_o$$
$$\hat{\mathbf{k}}_R = \hat{\mathbf{z}} \cos \theta_R + \hat{\mathbf{x}} \sin \theta_R. \qquad (4.5)$$

Thus we find for the phase of the wave u_o,

$$\varphi_o(x, z) = \mathbf{k}_o' \cdot \mathbf{r} = k'z \cos \theta_o + k'x \sin \theta_o \qquad (4.6)$$

and similarly for u_R,

$$\varphi_R(x, z) = \mathbf{k}_R' \cdot \mathbf{r} = k'z \cos \theta_R + k'x \sin \theta_R. \qquad (4.7)$$

The phase difference between the two waves as a function of position is given by

$$\delta(x, z) = \varphi_R(x, z) - \varphi_o(x, z)$$
$$= k'z(\cos \theta_R - \cos \theta_o) + k'x(\sin \theta_R - \sin \theta_o). \qquad (4.8)$$

A bright fringe is formed along the locus of all points for which $\delta(x, z) = 2m\pi$ with m an integer, or for

$$k'z(\cos \theta_R - \cos \theta_o) + k'x(\sin \theta_R - \sin \theta_o) = 2m\pi. \qquad (4.9)$$

Solving this for x, we obtain the equation defining the position of the fringes:

$$x = \left[\frac{\cos \theta_o - \cos \theta_R}{\sin \theta_R - \sin \theta_o} \right] z + \frac{m\lambda'}{\sin \theta_R - \sin \theta_o}. \qquad (4.10)$$

The coefficient of z defines the slope of the straight fringe; the second term on the right is the x-intercept. The integer m is the order of interference, which in general will depend on the phase difference between u_o and u_R at the point $(x, z) = (0, 0)$. The fringe spacing in the x-direction is the change in x corresponding to a change in m of one:

$$\Delta x = \frac{\lambda'}{\sin \theta_R - \sin \theta_o} \cdot \Delta m = \frac{\lambda'}{\sin_R \theta - \sin \theta_o} = d. \qquad (4.11)$$

The corresponding spatial frequency is

$$\nu_s = \frac{1}{d} = \frac{\sin \theta_R - \sin \theta_o}{\lambda'} \qquad (4.12)$$

This is the spatial frequency corresponding to a plane hologram. The radian spatial frequency in the x-direction is

$$\omega = 2\pi \nu_s = k'(\sin \theta_R - \sin \theta_o). \qquad (4.13)$$

In order to express ω as a function of the total angular separation between the two waves, define

$$\varphi \equiv 2\pi - \theta_o + \theta_R \qquad (4.14)$$

so that

$$\omega = k'(\sin \theta_R - \sin \theta_o)$$
$$= 2k' \sin \frac{\varphi}{2} \cos \left(\theta_R - \frac{\varphi}{2} \right). \qquad (4.15)$$

If α is the angle which the fringes make with the z-axis (see Fig. 4.2), then

$$\tan \alpha = \frac{\cos \theta_o - \cos \theta_R}{\sin \theta_R - \sin \theta_o} = \tan \left(\theta_R - \frac{\varphi}{2} \right) \qquad (4.16)$$

so that

$$\alpha = \theta_R - \frac{\varphi}{2} \qquad (4.17)$$

and

$$\omega = 2k' \sin \frac{\varphi}{2} \cos \alpha. \qquad (4.18)$$

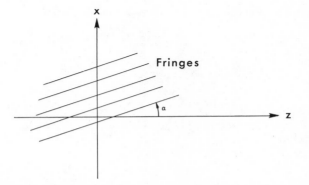

Fig. 4.2 Schematic of the fringe system in a thick recording medium.

This is the radian spatial frequency recorded in the x-direction for fringes produced by interference between two plane waves whose directions of propagation differ by an angle φ.

4.2 DIFFRACTION FROM A THREE-DIMENSIONAL GRATING

4.2.1 The Three-Dimensional Grating Equation

Now that the location and orientation of the fringes in the recording medium are known, we need to know how light is diffracted from such a three-dimensional grating. We begin with an approximation to the Fresnel-Kirchhoff diffraction integral. Referring to Fig. 4.3, we suppose that there is a line source at P_c which extends in the y-direction. The diffracting aperture lies in the x-y plane and we are assuming no y-variation so that the final integral will be two- instead of three-dimensional. The field at some point P_i located to the right of the aperture is given by

$$U(P_i) = A \cos \delta \left(\frac{-i}{\lambda}\right)^{\frac{1}{2}} \int_{\text{aperture}} \frac{e^{ik(r+s)}}{\sqrt{r+s}} \, dx. \qquad (4.19)$$

The quantity A is the amplitude of the cylindrical wave at unit distance from the line source P_c. If Q is a point in the aperture, then the distances r and s are $\overline{P_cQ}$ and $\overline{QP_i}$, respectively. The angle δ is the angle between the line P_cP_i and the z-axis. If the source P_c is located at (x_c, z_c) and P_i at (x_i, z_i) then

$$r_o^2 = x_c^2 + z_c^2$$
$$s_o^2 = x_i^2 + z_i^2, \qquad (4.20)$$

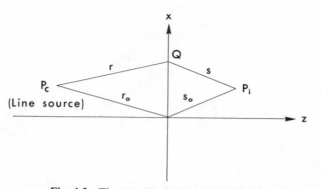

Fig. 4.3 The notation for the integral (4.19).

and if we assume that all distances are large compared with the extent of the aperture, we may write

$$r \sim r_o - \frac{x_c x}{r_o} + \frac{x^2}{2r_o} + \cdots$$

$$s \sim s_o - \frac{x_i x}{s_o} + \frac{x^2}{2s_o} + \cdots$$

(4.21)

by using the binomial expansion and ignoring the higher order terms. With this approximation (4.19) becomes

$$U(P_i) = A \cos \delta \left(\frac{-i}{\lambda r_o s_o} \right)^{\frac{1}{2}} e^{ik(r_o + s_o)}$$

$$\times \int_{\text{aperture}} \exp \left[-ik \left(\frac{x x_c}{r_o} - \frac{x^2}{2r_o} + \frac{x x_i}{s_o} - \frac{x^2}{2s_o} \right) \right] dx, \quad (4.22)$$

where we have made the approximations $(rs)^{-\frac{1}{2}} \sim (r_o s_o)^{-\frac{1}{2}}$. Define the direction cosines of the rays from P_c to Q and from Q to P_i as (Fig. 4.4)

$$l_c = -\frac{z_c}{r_o} \qquad l_i = \frac{z_i}{s_o}$$

$$m_c = -\frac{x_c}{r_o} \qquad m_i = \frac{x_i}{s_o}$$

(4.23)

so that

$$U(P_i) = A \cos \delta \left(\frac{-i}{\lambda r_o s_o} \right)^{\frac{1}{2}} e^{ik(r_o + s_o)}$$

$$\times \int_{\text{aperture}} \exp \left[-ik \left(x(m_i - m_c) - \frac{x^2}{2r_o} - \frac{x^2}{2s_o} \right) \right] dx. \quad (4.24)$$

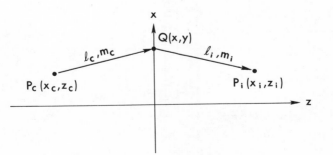

Fig. 4.4 Defining the direction cosines of the illuminating and diffracted waves.

As the distances $r_o, s_o \rightarrow \infty$, we may neglect the quadratic terms in the exponential under the integral. We assume that the factor A outside the integral tends to infinity along with r_o, s_o so that $U(P_i)$ does not vanish. This is the simpler case of Fraunhofer diffraction and we will consider only this. Hence we consider the case for $r_o, s_o \rightarrow \infty$, assuming that the terms in front of the integral (4.24) approach a constant C in the limit. We thus have

$$U(m_i) = C \int_{\text{aperture}} e^{-ikx(m_i - m_c)} \, dx \qquad (4.25)$$

as the basic equation governing Fraunhofer diffraction from a *thin* aperture.

In the three-dimensional case, we assume that the integral (4.25) yields the diffracted field in the direction m_i because of an elementary diffracting aperture of thickness dz located at $z = 0$. This is the usual "thin aperture" case. For an aperture with a nonnegligible thickness, we assume that we can just sum the contributions from each elementary thickness dz. This requires the assumption that the field diffracted from each dz is weak, so that the wave incident on each elementary aperture is the same. The solution will then be only a first-order approximation for the case of weak diffraction. Buckhardt [1] has found numerical solutions for the more rigorous case. Leith et al. [2] and Ramberg [3] essentially have solved the

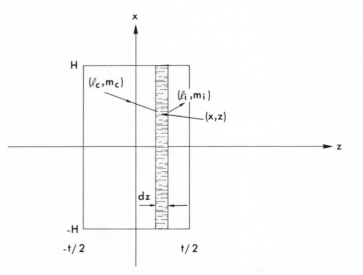

Fig. 4.5   Diffraction at a "thick aperture" as an extension of the usual "thin aperture" case.

weak diffraction problem. For a thick aperture, then, the contribution to the field in the direction $(l_i m_i)$ due to an elementary plane aperture dz at z is

$$d[U(l_i, m_i)] = \left[C \int_{\text{aperture}} e^{-ikx(m_i - m_c)} e^{-ikz(l_i - l_c)} dx \right] dz. \quad (4.26)$$

The geometry is shown in Fig. 4.5. The recording medium has a thickness t.

Because of the interference fringes within the aperture, the actual diffraction problem is not that of a clear aperture, but an aperture in which there is a spatial transmittance variation $G(x, z)$, called the "pupil function." Introducing the pupil function and summing all of the elementary contributions (4.26) over an aperture $2H$ and thickness t, we obtain

$$U(l_i, m_i) = C \int_{-H}^{H} \int_{-t/2}^{t/2} G(x, z) e^{-ikx(m_i - m_c)} e^{-ikz(l_1 - l_c)} dx \, dz. \quad (4.27)$$

4.2.2 The Pupil Function $G(x, z)$

Before we can evaluate the integral (4.27), we must find the form of the pupil function $G(x, z)$. To do this, we assume that the recorded signal of interest is proportional to the incident irradiance. Since the recorded signal of interest is actually the amplitude transmittance $t(x)$, the actual form is $t(x, z) = f(E_o) + \beta |H(x, z)|^2$, but the inclusion of the constant term $f(E_o)$ just adds an unnecessary complication at this point, so we will omit it. The irradiance produced by the interference of the two waves of Fig. 4.1 is

$$|H(x, z)|^2 = (u_o + u_R)(u_o^* + u_R^*) = a_o^2 + a_R^2 + 2a_o a \cos [\delta(x, z)], \quad (4.28)$$

where the phase difference function $\delta(x, z)$ is given by Eq. 4.8:

$$\delta(x, z) = k'z(\cos \theta_R - \cos \theta_o) + k'x(\sin \theta_R - \sin \theta_o).$$

Note that we are allowing for the possibility of different recording and illuminating wavelengths: the k' of Eq. 4.8 is the wave-number of the recording light in the medium, and k of Eq. 4.27 is the wave-number of the illuminating light in the medium. We now define two more sets of direction cosines, those of the reference and object waves (Fig. 4.6):

$$l_R = \cos \theta_R \qquad l_o = \cos \theta_o$$
$$m_R = \sin \theta_R \qquad m_o = \sin \theta_o \qquad (4.29)$$

so that we can write

$$\delta(x, z) = k'z(l_R - l_o) + k'x(m_R - m_o). \quad (4.30)$$

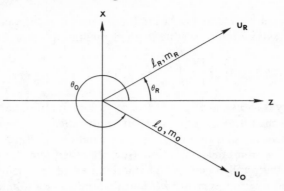

Fig. 4.6 Defining the direction cosines of object and reference waves. Both are assumed to be plane waves in this analysis.

Referring to Fig. 4.2, let the direction cosines of the fringes in the medium be given by l and m, where

$$l = \cos \alpha$$

$$m = \sin \alpha$$

(4.31)

so that

$$\tan \alpha = \frac{m}{l} = \frac{\cos \theta_o - \cos \theta_R}{\sin \theta_R - \sin \theta_o} = \frac{l_o - l_R}{m_R - m_o}.$$

(4.32)

Now (4.30) may be written

$$\delta(x, z) = \frac{k'}{l} (m_R - m_o)(lx - mz).$$

(4.33)

By using the relations (4.15) and (4.18) we can write

$$m_R - m_o = 2l \sin \frac{\varphi}{2}$$

(4.34)

so that

$$\delta(x, z) = 2k' \sin \frac{\varphi}{2} [lx - mz].$$

(4.35)

Hence we have

$$|H(x, z)|^2 = a_o^2 + a_R^2 + 2a_o a_R \cos \omega_o (lx - mz),$$

(4.36)

where we have defined

$$\omega_o = 2k' \sin \frac{\varphi}{2}.$$

(4.37)

Thus we can write the pupil function as

$$G(x, z) = E_o[1 + M \cos \omega_o(lx - mz)]$$

$$= E_o \left\{ 1 + \frac{M}{2} [e^{i\omega_o(lx-mz)} + e^{-i\omega_o(lx-mz)}] \right\} \qquad (4.38)$$

where E_o is the average exposure received by the recording medium and is proportional to $a_o^2 + a_R^2$. The quantity M is the exposure modulation and is proportional to $2a_o a_R/(a_o^2 + a_R^2)$.

4.2.3   The Bragg Condition

Substitution of (4.38) into (4.27) yields

$$U(l_i, m_i) = CE_o \int_{-H}^{H} \int_{-t/2}^{t/2} \left\{ e^{-ikx(m_i - m_c)} e^{-ikz(l_i - l_c)} + \frac{M}{2} [e^{i\omega_o(lx - mz)} e^{-ikx(m_i - m_c)} \right.$$

$$\left. \times e^{-ikz(l_i - l_c)} + e^{-i\omega_o(lx - mz)} e^{-ikx(m_i - m_c)} e^{-ikz(l_i - l_c)}] \right\} dx \, dz$$

which divides into three integrals

$$U(l_i, m_i) = \frac{MCE_o}{2} \int_{-H}^{H} \int_{-t/2}^{t/2} e^{-ix(km_i - km_c - \omega_o l)} e^{-iz(kl_i - kl_c + \omega_o m)} \, dx \, dz$$

$$+ \frac{MCE_o}{2} \int_{-H}^{H} \int_{-t/2}^{t/2} e^{-ix(km_i - km_c + \omega_o l)} e^{-iz(kl_i - kl_c - \omega_o m)} \, dx \, dz$$

$$+ CE_o \int_{-H}^{H} \int_{-t/2}^{t/2} e^{-ikx(m_i - m_c)} e^{-ikz(l_i - l_c)} \, dx \, dz. \qquad (4.39)$$

Since

$$\int_{-H}^{H} e^{-iBx} \, dx = 2H \frac{\sin BH}{BH} = 2H \operatorname{sinc} BH,$$

Equation 4.39 is readily integrated to

$$U(l_i, m_i) = \frac{MCE_o}{2} \left\{ \left[t \operatorname{sinc} [(kl_i - kl_c + \omega_o m) \frac{t}{2}] \right] \right.$$

$$\times \left[2H \operatorname{sinc} [(km_i - km_c - \omega_o l)H] \right] \right\}$$

$$+ \frac{MCE_o}{2} \left\{ \left[t \operatorname{sinc} [(kl_i - kl_c - \omega_o m) \frac{t}{2}] \right] \right.$$

$$\times \left[2H \operatorname{sinc} [(km_i - km_c + \omega_o l)H] \right] \right\}$$

$$+ 2CE_o Ht \left\{ \operatorname{sinc} [(kl_i - kl_c) \frac{t}{2}] \right\}$$

$$\times \left\{ \operatorname{sinc} [(km_i - km_c)H] \right\}. \qquad (4.40)$$

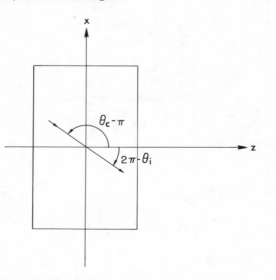

Fig. 4.7 The directly transmitted zero-order wave.

This equation is the first-order solution to the problem of diffraction from a three-dimensional sinusoidal grating. The last term is a maximum when

$$l_i = l_c \qquad \text{and} \quad m_i = m_i. \tag{4.41}$$

Since $l_c^2 + m_c^2 = 1$ and $l_i^2 + m_i^2 = 1$, the fact that $l_i = l_c$ implies $m_i = m_c$. This condition is illustrated in Fig. 4.7. Since we are using the Fraunhofer

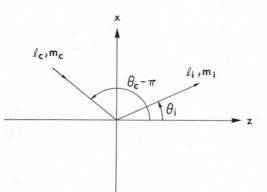

Fig. 4.8 Illustrating the angles and direction cosines for the illuminating and diffracted waves.

approximation, l_c and l_i, as defined by (4.23), now refer to the angles θ_c and θ_i shown in Fig. 4.8, that is,

$$l_c = \cos \theta_c \quad \text{and} \quad m_c = \sin \theta_c$$
$$l_i = \cos \theta_i \qquad m_i = \sin \theta_i. \tag{4.42}$$

This is just the directly transmitted or zero-order wave.

The first two terms in (4.40) are just the two first-order diffracted waves, corresponding to the primary and conjugate images. The first term is a maximum when the arguments of both sinc functions are zero, or when

$$l_i = l_c - \frac{\omega_o m}{k} \quad \text{and} \quad m_i = m_c + \frac{\omega_o l}{k}. \tag{4.43}$$

We can write the grating equation as

$$d(\sin \theta_i - \sin \theta_c) = \lambda \tag{4.44}$$

for the first diffracted order. Since $m_c = \sin \theta_c$ and $m_i = \sin \theta_i$, the grating equation may be written

$$m_i - m_c = \frac{\lambda}{d} = \frac{2\pi \nu_s}{k} = \frac{\omega}{k} \tag{4.45}$$

where ω is the radian spatial frequency of the grating in the x-direction. From (4.18), (4.32), and (4.37), we have

$$\omega = \omega_o l \tag{4.46}$$

and so the grating equation becomes

$$m_i - m_c = \frac{\omega_o l}{k}. \tag{4.47}$$

Hence we see that the second condition of (4.43) implies simply that there is a diffraction maximum in the direction predicted by the grating equation.

Referring to Fig. 4.9, we see that we can write the law of reflection in the form

$$l_c l + m_c m = l_i l + m_i m \tag{4.48}$$

since $\cos i = l_c l + m_i m$ and $\cos r = l_i l + m_i m$.

Hence the law of reflection can be written as

$$l_c - l_i = (m_i - m_c)\frac{m}{l} \tag{4.49}$$

for an incident ray with direction cosines l_c, m_c being reflected from a plane with direction cosines l, m into the direction $l_i m_i$. Since the first-order

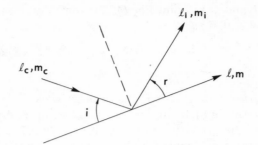

Fig. 4.9 Illustrating how the reflection law ($r = i$) can be expressed in terms of the direction cosines.

diffracted wave is in the direction required by (4.47), Eq. 4.49 may be written

$$l_c - l_i = \frac{\omega_o m}{k} \tag{4.50}$$

which is just the first of the conditions (4.43). Hence we see that there is a maximum for the first term of (4.40) in the direction l_i, m_i, such that the

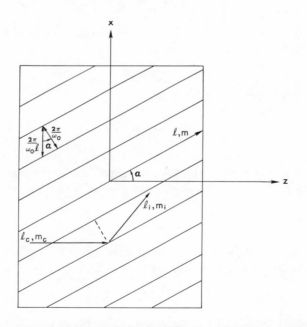

Fig. 4.10 Illustrating the Bragg condition for diffraction. The entering wave in the direction (l_c, m_o) is diffracted into the direction (l_i, m_i) by the fringes lying in the direction (l, m).

grating equation is satisfied simultaneously with an apparent reflection from the plane of the fringes lying in the l, m direction. These two conditions constitute the *Bragg condition* for diffraction from a three-dimensional grating. This is illustrated in Fig. 4.10. Similar arguments apply to the second term of (4.40). Note that it is impossible to have simultaneous maxima in both first orders. The first term in (4.40) is maximum when

$$\frac{-(l_i - l_c)}{m} = \frac{\omega_o}{k}$$

and $\hspace{6cm}$ (4.51)

$$\frac{m_i - m_c}{l} = \frac{\omega_o}{k}$$

and the second term is a maximum when

$$\frac{l_i - l_c}{m} = \frac{\omega_o}{k}$$

and $\hspace{6cm}$ (4.52)

$$\frac{-(m_i - m_c)}{l} = \frac{\omega_o}{k}.$$

The directions implied by (4.52) are just the reverse of those in (4.51).

4.2.4 Wavelength Change

To see the effect of using different wavelengths in recording and illuminating, we write (4.51) as

$$\frac{-(l_i - l_c)}{m} = 2\frac{k'}{k}\sin\frac{\varphi}{2} = \frac{m_i - m_c}{l} \qquad (4.53)$$

where we have used $\omega_o = 2k' \sin(\varphi/2)$ [cf. (4.37)]. Recall that $k' = 2\pi/\lambda'$, where λ' is the wavelength of the recording light in the medium, and $k = 2\pi/\lambda$, λ being the illuminating wavelength. The angle φ is the total angle between the two beams used to record the grating. The Bragg condition thus becomes

$$\frac{l_c - l_i}{m} = 2\frac{\lambda}{\lambda'}\sin\frac{\varphi}{2}$$

and $\hspace{6cm}$ (4.54)

$$\frac{m_i - m_c}{l} = 2\frac{\lambda}{\lambda'}\sin\frac{\varphi}{2}.$$

But since $m_R - m_o = 2l \sin(\varphi/2)$ (4.34) and $l_R - l_o = -2m \sin(\varphi/2)$, we can write the Bragg condition in terms of the direction cosines of the recording beams as

$$m_i - m_c = \frac{\lambda}{\lambda'}(m_R - m_o)$$

and   (4.55)

$$l_i - l_c = \frac{\lambda}{\lambda'}(l_R - l_o).$$

From these equations we can readily see that for $\lambda = \lambda'$ and $l_c = -l_R$ (illuminating beam reversed in direction from the reference wave), $l_i = -l_o$, that is, the diffracted wave is traveling in a direction opposite to the object wave along the same path. This is the conjugate image term.

Using (4.52) we obtain a set of equations similar to (4.55) for the primary wave

$$m_i - m_c = -\frac{\lambda}{\lambda'}(m_R - m_o)$$

                                                                                            (4.56)

$$l_i - l_c = -\frac{\lambda}{\lambda'}(l_R - l_o).$$

Here we see that for $\lambda = \lambda'$ and $l_c = l_R$ (illuminating beam identical to the reference beam), $l_i = l_o$, that is, the diffracted wave is identical to the original object wave.

We may generalize to the case of a more complicated object by noting that the first term in (4.40) will give rise to a conjugate (real) image when all of the rays of the reference beam are reversed in direction and the same wavelength is used in recording and illuminating. Thus if the reference wave is a diverging spherical wave, a real image will be formed of the object when the illuminating wave is spherical and converges to the same point in space from which the reference wave diverged.

The second term of (4.40) will yield a primary (virtual) image in the same position as the object, if the illuminating wave is identical to the reference wave.

There is further solution to (4.55) and (4.56) for $l_c = l_o$ and $l_c = -l_o$, respectively. Either choice of illuminating wave then leads to $l_i = l_R$ or $l_i = -l_R$. This is a valid solution (for $\lambda = \lambda'$) for the case of a plane object wave as considered here. For a more complicated object wave, however, the illuminating wave will not in general match exactly the reference wave. The condition $l_c = l_o$ is then only satisfied for a single component of the plane wave spectrum of the object wave and no complete image will be formed.

If the wavelength of the illuminating wave differs from that of the recording waves ($\lambda \neq \lambda'$), it is still possible to find solutions to (4.55) and (4.56). When we make the generalization from a single plane object wave to a more general object wave represented by a spectrum of plane waves, however, we see that the Bragg condition cannot be satisfied simultaneously by all components. We can see this by differentiating (4.55); for example,

$$\Delta l_c = \frac{\Delta \lambda}{\lambda'} (l_R - l_o). \tag{4.57}$$

Here we see that the change in direction Δl_c of the illuminating beam required by a change in wavelength $\Delta \lambda$ is a function of the direction of the object wave l_o. Hence no single illuminating wave can simultaneously satisfy the Bragg condition for a whole spectrum of object waves. In general, no complete image will be formed with a change in wavelength.

A further restriction on the allowable change in wavelength results when we require that no light be diffracted at angles greater than 90°. Hence from (4.45) we must have the grating spacing $d \geq \lambda$. Writing d as $2\pi/\omega$ with ω given by (4.18), we obtain the requirement

$$\frac{\lambda}{\lambda'} \leq \frac{1}{m_R - m_o}. \tag{4.58}$$

4.2.5 Orientation Sensitivity

All of the foregoing conclusions apply only for truly thick holograms where the fringe spacing is very much smaller than the thickness of the recording medium. For most holograms made on photographic emulsions this is only approximately true, and there generally will be a finite range of directions over which an object wave can be reconstructed. To determine just how large this range is, we return to (4.40). If we consider only the second term, we see that the amplitude diffracted into the first order is

$$U(l_i, m_i) = MCE_oHt \left\{ \text{sinc} \left[(kl_i - kl_c - \omega_o m) \frac{t}{2} \right] \right\}$$

$$\times \left\{ \text{sinc} \left[(km_i - km_c + \omega_o l)H \right] \right\} \tag{4.59}$$

where t is the thickness of the hologram and H its dimension. We now construct a ratio I/I_o, in which

$$I_o = [MCE_oHt]^2 \tag{4.60}$$

is the irradiance of the diffracted wave when the Bragg condition is satisfied, and

$$I = I_o \operatorname{sinc}^2 \left[(kl_i - kl_c - \omega_o m) \frac{t}{2} \right] \tag{4.61}$$

is the irradiance of the diffracted wave in a direction determined by the grating equation, but the incident and diffracted rays do not appear to be reflected from the Bragg planes. We have then

$$\frac{I}{I_o} = \operatorname{sinc}^2 \left[(kl_i - kl_c - \omega_o m) \frac{t}{2} \right]$$

$$= \operatorname{sinc}^2 \left\{ [k(l_i - l_c) + k'(l_R - l_o)] \frac{t}{2} \right\} . \tag{4.62}$$

When the argument of the sinc function in (4.62) equals π, the relative irradiance in the diffracted beam will be zero. This indicates, for example, the minimum angular rotation of the hologram which will extinguish the image. Alternatively, it indicates the minimum angular displacement of the reference beam in order to record more than one hologram which can be read out separately without overlapping images.

Let us first assume that $\lambda = \lambda'$, so that the argument of the sinc function becomes

$$(l_i - l_c + l_R - l_o) \frac{\pi t}{\lambda} . \tag{4.63}$$

A small rotation of the hologram results in a change Δl_c in l_c with the corresponding change in direction of the diffracted wave Δl_i in l_i. The argument of the sinc function now becomes

$$(l_i + \Delta l_i - l_c - \Delta l_c + l_R - l_o) \frac{\pi t}{\lambda} \tag{4.64}$$

and when this equals π, I/I_o becomes zero:

$$(l_i + \Delta l_i - l_c - \Delta l_c + l_R - l_o) \frac{\pi t}{\lambda} = \pi. \tag{4.65}$$

If we assume that we started with the illuminating wave identical to the reference wave ($l_c = l_R$), and therefore the diffracted wave identical to the object wave ($l_i = l_o$), Eq. 4.65 becomes

$$\Delta l_i - \Delta l_c = \frac{\lambda}{t} . \tag{4.66}$$

Similarly, since we are assuming that the grating equation is satisfied, the argument of the second sinc function of (4.59) remains zero and therefore we must have

$$\Delta m_i - \Delta m_c = 0. \tag{4.67}$$

Now since

$$\Delta l_i = -\sin \theta_i \, \Delta\theta_i = -m_i \, \Delta\theta_i$$

$$\Delta l_c = -\sin \theta_c \, \Delta\theta_c = -m_c \, \Delta\theta_c$$

$$\Delta m_i = \cos \theta_i \, \Delta\theta_i = l_i \, \Delta\theta_i$$

$$\Delta m_c = \cos \theta_c \, \Delta\theta_c = l_c \, \Delta\theta_c,$$

(4.68)

Equation 4.67 gives

$$\Delta\theta_i = \frac{l_c}{l_i} \Delta\theta_c.$$

(4.69)

Substitution into (4.66) then yields

$$\Delta\theta_c = \frac{\lambda}{t} \left(m_c - \frac{m_i l_c}{l_i} \right)^{-1}$$

(4.70)

and since $(l_i, m_i) = (l_o, m_o)$ and $(l_c, m_c) = (l_R, m_R)$, we have

$$\Delta\theta_c = \frac{\lambda}{t} \left(\frac{l_o}{m_R l_o - m_o l_R} \right)$$

(4.71)

with a similar expression for the image shift

$$\Delta\theta_i = \frac{\lambda}{t} \left(\frac{l_R}{m_R l_o - m_o l_R} \right).$$

(4.72)

Equation 4.71 gives the angular rotation in radians required to extinguish the image formed with a hologram of thickness t and made and illuminated with light of wavelength λ *in the medium*. The hologram was recorded with two plane waves with direction cosines (l_o, m_o) and (l_R, m_R). When $(l_o, m_o) = (1, 0)$ (object wave incident normally), Eq. 4.71 becomes

$$\Delta\theta_c = \frac{\lambda}{t} \frac{1}{m_R} = \frac{\lambda}{t} \csc \theta_R.$$

(4.73)

Figure 4.11 shows a plot of this function for several values of λ/t. The values of both λ and θ_R refer to those in the recording medium. Not all angles θ_R are readily accessible, because an angle of incidence of 90° outside the recording medium results in an angle $\theta_R \cong 40°$ inside for an index of 1.5. Angles greater than 90° correspond to the reference wave entering from the back side of the hologram. Figure 4.12 indicates the conditions assumed for calculating the curves of Fig. 4.11. In 4.12a the hologram is recorded with the (plane) object wave incident normally, so that $\theta_o = 0$ and $(l_o, m_o) = (1, 0)$. The reference wave makes an angle θ_R with the z-axis. Fringes are recorded lying in the direction (l, m) which is the bisector of the angle between the object and reference waves. The radian spatial frequency ω of the recorded grating depends on the angle between these two waves:

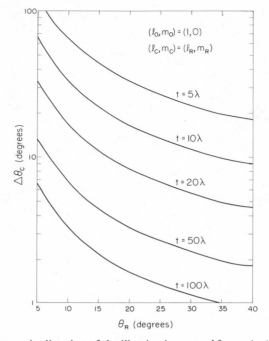

Fig. 4.11   The change in direction of the illuminating wave $\Delta\theta_c$ required to extinguish the diffracted wave as a function of the angle of incidence of the reference wave θ_R. The curves are plotted for $\theta_o = 0$ and for several values of hologram thickness. The wavelength and angles refer to values in the medium.

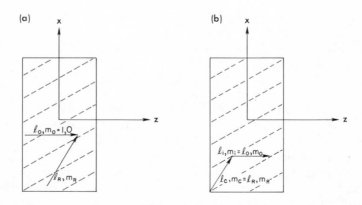

Fig. 4.12   Illustrating the assumed recording and illuminating conditions for the curves of Fig. 4.11. (a) Recording. (b) Illuminating.

$$\omega = \frac{2\pi}{\lambda}(\sin\theta_R - \sin\theta_o)$$

$$= \frac{2\pi}{\lambda}\sin\theta_R \qquad \text{for} \quad \lambda = \lambda', \theta_o = 0.$$

Hence small values of θ_R imply low spatial frequencies. Figure 4.12b shows that we are also assuming that the grating is illuminated with a plane wave in the same direction as the reference wave: $(l_c, m_c) = (l_R, m_R)$. The diffracted wave is then in the same direction as the object wave: $(l_i, m_i) = (l_o, m_o)$.

Figure 4.11 shows that for small θ_R, a relatively large change in θ_c is required to extinguish the diffracted wave. Hence the irradiance of the diffracted wave is relatively insensitive to the orientation of the hologram (θ_c can be changed by rotating the hologram) if the reference wave makes only a small angle with the object wave. The curves of Fig. 4.11 are symmetrical about $\theta_R = 90°$, so the irradiance of the diffracted wave is also relatively insensitive to orientation if the reference wave is incident from the back side of the hologram. This means that the light is incident in a direction nearly opposite that of the object beam. These are called reflection or Lippmann holograms, and are described in Section 4.3. On the other hand, the amount of flux diffracted is very sensitive to hologram orientation when the object and reference waves meet at a large angle, and therefore the recorded fringes are very closely spaced.

4.2.6 Wavelength Sensitivity

Next, let us consider the effect of illuminating the hologram with light of a wavelength different from that used for recording. In (4.59) we see that the diffracted amplitude corresponding to the primary wave is proportional to the product of two sinc functions of arguments

$$(kl_i - kl_c - \omega_o m)\frac{t}{2} \tag{4.74}$$

and

$$(km_i - km_c + \omega_o l)H. \tag{4.75}$$

From (4.37) we obtain

$$\omega_o = 2k'\sin\frac{\varphi}{2} \tag{4.76}$$

and from (4.35) we obtain

$$2l \sin \frac{\varphi}{2} = m_R - m_o$$

(4.77)

$$-2m \sin \frac{\varphi}{2} = l_R - l_o$$

so that (4.74) and (4.75) may be written as

$$[k(l_i - l_c) + k'(l_R - l_o)] \frac{t}{2}$$

(4.78)

and

$$[k(m_i - m_c) + k'(m_R - m_o)]H.$$

(4.79)

Assume that we start with $\lambda = \lambda'$ and the condition for a maximum for which $(l_i, m_i) = (l_o, m_o)$ and $(l_c, m_c) = (l_R, m_R)$. We now ask what change in wavelength, with the illuminating wave in the same direction as the reference wave $(l_c = l_R)$ will cause Eq. 4.62 to vanish? This will occur when the argument of the sinc function (4.78) equals π. A change in wavelength results in a change in direction of the diffracted wave, so that as $k' \rightarrow k + \Delta k$, $l_i \rightarrow l_i + \Delta l_i$. We then have as the condition for a zero

$$[k(l_i + \Delta l_i - l_c) + k(l_R - l_o) + \Delta k(l_R - l_o)] \frac{t}{2} = \pi$$

(4.80)

and

$$[k(m_i + \Delta m_i - m_c) + k(m_R - m_o) + \Delta k(m_R - m_o)]H = \pi.$$

(4.81)

But since we have chosen the condition $(l_i, m_i) = (l_o, m_o)$ and $(l_c, m_c) = (l_R, m_R)$, these reduce to

$$\Delta k(l_R - l_o) + k \Delta l_i = \frac{2\pi}{t}$$

(4.82)

and

$$\Delta k(m_R - m_o) + k\Delta m_i = 0$$

(4.83)

where $k = 2\pi/\lambda$ and $\Delta k = -(2\pi/\lambda^2) \Delta\lambda$. Using (4.68) we obtain the change in wavelength required for extinction of the image:

$$\Delta\lambda = \frac{\lambda^2}{t} \left(\frac{l_o}{1 - l_o l_R - m_o m_R} \right).$$

(4.84)

This change in wavelength corresponds to a change in direction of the diffracted beam given by

$$\Delta\theta_i = \frac{\lambda}{t} \left(\frac{m_R - m_o}{1 - l_o l_R - m_o m_R} \right).$$

(4.85)

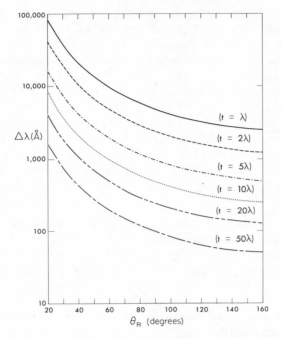

Fig. 4.13 The change in wavelength of the illuminating wave $\Delta\lambda$ required to extinguish the diffracted wave as a function of the angle of incidence of the reference wave θ_R. The curves are plotted for $\theta_o = 0$ and for several values of hologram thickness. The wavelength and angles refer to values in the medium.

Figure 4.13 shows curves of $\Delta\lambda$ for several values of λ/t (in the medium) for the case $(l_o, m_o) = (1, 0)$.

On the basis of Figs. 4.11 and 4.13 we can, following Leith et al. [2], classify holograms into three general categories, depending on the value of φ, the total angle between reference and object waves.

Class A: φ is small, so that ω_o is small compared to $1/t$. In this case the diffracted irradiance is relatively insensitive to change in illuminating wavelength or direction and the recording medium behaves as a two-dimensional medium.

Class B: φ is moderately large (10 to 120° for photographic emulsions). In this region the diffracted irradiance is most sensitive to change in the direction of the illuminating wave, so that hologram alignment is critical. A number of different holograms may be stored on a single photographic plate, each of which may be read out without interference from the others, if each is made with the reference beam in a different direction. The diffracted irradiance is fairly sensitive to changes in illuminating wavelength and one can suppress wavelengths separated by several hundred angstroms from the recording wavelength with most photographic emulsions.

Class C: φ is large, with object and reference waves entering from opposite sides of the recording medium. In this case the recorded fringes (Bragg planes) lie nearly parallel to the *x-y* plane of the hologram. This type of hologram is most sensitive to a change in wavelength; the fringes are separated by only about one-half the wavelength in the medium. Holograms of this class may be illuminated with white light, since it will act as an interference filter, producing a diffracted wave for only a narrow band of wavelengths.

This type of hologram, however, is relatively insensitive to a change in direction of the illuminating wave. A rotation of the hologram does not cause the extinction of the image but rather a change in color so as to maintain the Bragg condition (Eqs. 4.55 and 4.56). This class of hologram was first described by Denisyuk [4], who noted its similarity to Lippmann color photography. This type of hologram is sometimes called a Lippmann or reflection hologram, and because of its importance we shall describe it in detail in the next section.

4.3   REFLECTION HOLOGRAMS [4]

Reflection holograms are made by allowing the reference and object beams to enter the recording medium from opposite sides, so that they are traveling in approximately opposite directions (Fig. 4.14). The interference of these two beams forms a stationary standing wave pattern in the recording medium. After processing, an image exists in the recording medium with a density which is proportional to the irradiance of the standing wave pattern. The fringes formed are approximately planes perpendicular to the *z*-axis. If such a hologram is illuminated with a beam of light similar to the reference beam, it will reflect a portion of this light which is identical to the original object wave. The reconstruction is viewed in reflection rather than in transmission as with the usual hologram. White light may also be used for reconstruction, since the wavelength sensitivity of this type of hologram is very high (Fig. 4.13).

To consider the theory of the process, we write the object and reference waves as $O_o(x, y, z)e^{i\varphi_o(x,y,z)}$ and $R_o(x, y, z)e^{i\varphi_R(x,y,z)}$, respectively. The total field in the volume of the recording medium is

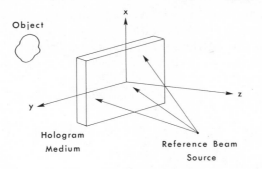

Fig. 4.14 Illustrating the general arrangement for recording a reflection hologram.

$$H(x, y, z) = O_o e^{i\varphi_o} + R_o e^{i\varphi_R}$$

and the resultant irradiance of the standing wave pattern is

$$|H|^2 = O_o^2 + R_o^2 + 2O_o R_o \cos(\varphi_o - \varphi_R). \tag{4.86}$$

The fringes are the loci of constant phase difference

$$\delta = \varphi_o - \varphi_R \tag{4.87}$$

so that we can write for a fringe

$$|H|^2 = O_o^2 + R_o^2 + 2O_o R_o \cos\delta. \tag{4.88}$$

Now assume that the recording medium is such that the resultant recorded signal D is proportional to $|H|^2$:

$$D = \chi|H|^2. \tag{4.89}$$

Assuming that the dielectric constant ϵ of the medium is related to D, we can, for small D, expand $\epsilon(D)$ into a series

$$\epsilon(D) = \epsilon|_{D=0} + D\frac{\partial\epsilon}{\partial D}\Big|_{D=0} + \cdots$$

$$= \epsilon_o + \gamma D \tag{4.90}$$

where $\epsilon|_{D=0}$ is the dielectric constant of the unexposed recording medium. Substituting (4.88) into (4.89), and then substituting this value of D into (4.90), we obtain for the dielectric constant

$$\epsilon = \epsilon_o + \gamma\chi(O_o^2 + R_o^2 + 2O_o R_o \cos\delta). \tag{4.91}$$

Let us separate the infinitely thin layer enclosed between the isophase surface described by $\delta = $ constant and the isophase surface described by

$\delta + d\delta$ = constant. Such a layer may be regarded as the interface between two media, where $d\epsilon$ equals the difference between the dielectric constants of these media. This interface will then behave as a mirror surface. Writing the amplitude reflection coefficient as

$$dr = \frac{\sqrt{(\epsilon + d\epsilon)/\epsilon} - 1}{\sqrt{(\epsilon + d\epsilon)/\epsilon} + 1} \qquad (4.92)$$

and using the binomial expansion for the square roots, we find

$$dr \cong \frac{d\epsilon}{4\epsilon}, \qquad (4.93)$$

with $d\epsilon = 2\gamma\chi O_o R_o \sin \delta \, d\delta$. Hence we can write for the amplitude reflectance of this layer

$$dr = CO_o \sin \delta \, d\delta \qquad (4.94)$$

where

$$C = \frac{\gamma\chi R_o}{2\epsilon}. \qquad (4.95)$$

Thus the amplitude reflected from this layer is proportional to the amplitude of the light scattered by the object. Let us now see how a beam of light interacts with this layer. Let a beam of light identical to the reference beam ($R_o e^{i\varphi_R}$) be incident on the hologram. Following Denisyuk, we take advantage of an assumed low density of the image in the recording medium and determine the form of the radiation reflected from each individual isophase layer in order to combine these later to determine the form of the radiation reflected by the entire hologram. Multiplying the incident wave $R_o e^{i\varphi_R}$ by the amplitude reflectance dr, we find the amplitude of the wave at the surface of the layer:

$$d\psi = CR_o e^{i\varphi_R} O_o \sin \delta \, d\delta. \qquad (4.96)$$

But from Eq. 4.87, $\varphi_R = \varphi_o - \delta$ on the isophase surface, so that

$$d\psi = (CR_o \sin \delta \, d\delta e^{-i\delta}) O_o e^{i\varphi_o}. \qquad (4.97)$$

We see that the wave reflected from the isophase layer is identical to the original object wave except for the constant multiplier in brackets in (4.97). An observer viewing this wave will see a three-dimensional virtual image of the object. A real image is formed when the illuminating wave converges to a real image of the reference beam source.

By summing the waves reflected by all of the elementary isophase layers, it is possible to show that this type of hologram also reproduces the spectral composition of the object wave. Thus the object wave can be reconstructed even if the hologram is illuminated with white light. To show this, the illuminating wave is written as

$$\int_0^\infty R(k')e^{ik'\mathcal{L}(\mathbf{r})}\,dk' \tag{4.98}$$

where $\mathcal{L}(\mathbf{r})$ describes the optical path of the light and satisfies the eikonel equation

$$[\nabla\mathcal{L}(\mathbf{r})]^2 = n^2(\mathbf{r}) \tag{4.99}$$

and $\mathbf{r}$ is the position vector. Substituting (4.98) into (4.96) in place of $R_o e^{i\varphi_R}$ we have

$$d\psi = CO_o \sin\delta\, d\delta \int_0^\infty R(k')e^{ik'\mathcal{L}(\mathbf{r})}\,dk'. \tag{4.100}$$

But

$$\delta = \varphi_o - \varphi_R = k\mathcal{L}_o(\mathbf{r}) - k\mathcal{L}(\mathbf{r}) = k\rho \tag{4.101}$$

where $\mathcal{L}_0(\mathbf{r})$ describes the optical path of the object beam and $\mathcal{L}(\mathbf{r})$ that of the reference beam. We are assuming that the illuminating beam (4.98) follows the same path as the reference beam. The quantity ρ is the optical path difference between the two beams and k is $2\pi/\lambda$. Using (4.101) we can write

$$d\psi = kCO_o \sin\rho\, d\rho \int_0^\infty R(k')e^{ik\mathcal{L}_o(\mathbf{r})}e^{-ik'\rho}\,dk'. \tag{4.102}$$

Summing the contributions from each isophase layer between the layers described by ρ_1 and ρ_2 (assuming weak reflection), the reflected wave of interest becomes

$$\psi = kCO_o \int_{\rho_\cdot}^{\rho_2} \frac{e^{ik\rho} - e^{-ik\rho}}{2i}\,d\rho \int_0^\infty R(k')e^{ik'\mathcal{L}_o(\mathbf{r})}e^{-ik'\rho}dk'. \tag{4.103}$$

Changing the order of integration, this becomes

$$\psi = \frac{-ikCO_o}{2}\int_0^\infty R(k')e^{ik'\mathcal{L}_o(\mathbf{r})}\,dk'\int_{\rho_1}^{\rho_2}[e^{i\rho(k-k')} - e^{-i\rho(k+k')}]\,d\rho. \tag{4.104}$$

Carrying out the ρ-integration we obtain

$$\psi = \frac{-ikCO_o}{2}\int_0^\infty R(k')e^{ik'\mathcal{L}_o(\mathbf{r})}[\delta_1(k+k') - \delta_1(k-k')]\,dk' \tag{4.105}$$

where

$$\delta_1(k - k') = \frac{e^{i\rho_2(k-k')} - e^{i\rho_1(k-k')}}{i(k - k')} \tag{4.106}$$

which can be written

$$\delta_1(k - k') = \frac{2e^{(i/2)(\rho_1+\rho_2)(k-k')}}{k - k'} \sin\left[\frac{(\rho_2 - \rho_1)(k - k')}{2}\right]. \tag{4.107}$$

The functions $\delta_1(k + k')$ and $\delta_1(k - k')$ have the properties of Dirac delta functions. The function $\delta_1(k - k')$ has a maximum for $k = k'$, where it assumes the value $\rho_2 - \rho_1$. Similarly, $\delta_1(k + k')$ has its maximum for $k = -k'$. The widths of these functions are given by

$$\Delta k = \frac{2\pi}{\rho_2 - \rho_1}. \tag{4.108}$$

Since the thickness of the hologram is assumed much larger than a wavelength, $\rho_2 - \rho_1 \gg 1/k$, and the delta functions may be replaced by rectangular pulses of width Δk and height $\rho_2 - \rho_1$. Making this substitution, we find

$$\psi(\mathbf{r}) = \frac{-ikCO_o(\mathbf{r})}{2} (\rho_2 - \rho_1) \int_{k-(\Delta k)/2}^{k+(\Delta k)/2} R(k')e^{ik'\mathcal{L}_o(\mathbf{r})}dk'. \tag{4.109}$$

Therefore we see that the reflected wave is composed of wavelengths that differ little from the wavelength which exposed the hologram: its phase agrees with the phase of the wave scattered by the object; its amplitude is proportional to the amplitude of the object wave and the optical path $\rho_2 - \rho_1$ which the reconstructed wave travels through the hologram.

Making the appropriate changes of sign, we may also show that a conjugate image of the object is formed. The zero-order image is the light which is specularly reflected from the boundary of the hologram medium.

Thus we see that a hologram made in this manner has some very interesting properties, the most noteworthy being that white-light reconstruction is possible. This was expected because of the great wavelength sensitivity shown in Fig. 4.13. Pennington and Lin [5] have used this effect to produce a color hologram. This was done by using a reference beam composed of red and blue light. A color transparency was illuminated by the same combination. When the resulting hologram was illuminated with white light, a good two-color reproduction of the transparency was obtained. The method has also been extended to three colors, and color holograms represent a large portion of present-day research in holography.

REFERENCES

[1] C. B. Buckhardt, *J. Opt. Soc. Am.*, **56,** 1502 (1966).
[2] E. N. Leith, A. Kozma, J. Upatnieks, J. Marks, and N. Massey, *Appl. Opt.*, **5,** 1303 (1966).
[3] E. G. Ramberg, *R.C.A. Rev.*, **27,** 467 (1966).
[4] Y. N. Denisyuk, *Optics and Spectroscopy*, **15,** 279 (1963).
[5] K. S. Pennington and L. H. Lin, *Appl. Phys. Letters*, **7,** 56 (1965).

5 Factors Affecting Image Resolution

5.0 INTRODUCTION

In this chapter we will examine several of the factors that can limit the resolution in a holographic image—the size and bandwidth of the source (or sources) providing the reference and illuminating beams, the resolution capabilities of the recording medium, and the geometrical aberrations that can arise. Each of these factors is treated separately, as if it alone were degrading the image, although in actual practice, of course, they will be acting simultaneously. The ultimate, diffraction-limited resolution is derived as a natural consequence of the treatment in Section 5.1. No actual numbers or statements of the realizable resolution are given in Section 5.4, since the actual final image resolution is such a complicated function of the aberrations of the system.

5.1 REFERENCE AND ILLUMINATING BEAM SOURCE SIZE

In all of the analyses presented so far, we have assumed that the reference beam originated at a point source, resulting in a strictly plane or spherical reference beam. The same assumption has been made regarding the illuminating beam. In any real case, however, one never has a true mathematical point source; there is always some finite size associated with the reference and illuminating beam sources. Also, in many cases the two sources may be physically different, such as when a change in wavelength or a change in divergence is desired. In this section we will determine the effect of nonzero source size on the holographic image. As usual, we will

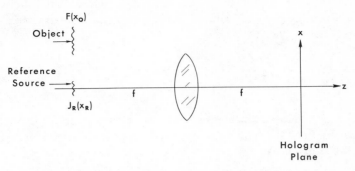

Fig. 5.1 A Fourier transform system. The reference beam is derived from a nonpoint source described by $J_R(x_R)$.

treat the problem in two dimensions only, since consideration of the three-dimensional problem only adds to the length of the equations and not to the physical insight.

Consider the Fourier transform system of Fig. 5.1. We assume that the lens is large enough so that we can neglect the effects of vignetting. The complex amplitude distribution across the object is represented by $F(x_o)$ and across the reference source by $J_R(x_R)$. Both the object and the reference source are in the same plane, a distance f from the lens of focal length f. The hologram plane is the x plane and is located a distance f behind the lens. The object field distribution in this plane is [see (A.18)]

$$O(x) = \left(\frac{-i}{\lambda f}\right)^{1/2} \int_{-\infty}^{\infty} F(x_o)e^{-i(k/f)xx_o}\,dx_o. \tag{5.1}$$

Similarly, the reference field at the hologram is given by

$$R(x) = \left(\frac{-i}{\lambda f}\right)^{1/2} \int_{-\infty}^{\infty} J_R(x_R)e^{-i(k/f)xx_R}\,dx_R. \tag{5.2}$$

The exposure term leading to the primary image wave is OR^*, and we will consider this term alone. Assuming simple equality between exposure and final amplitude transmittance, illuminating the hologram with a wave $C(x)$ yields a transmitted wave

$$\psi(x) = C(x)O(x)R^*(x). \tag{5.3}$$

We write the illuminating wave in a manner similar to (5.2) for the reference wave:

$$C(x) = \left(\frac{-i}{\lambda f}\right)^{1/2} \int_{-\infty}^{\infty} J_c(x_c)e^{-i(k/f)xx_c}\,dx_c, \tag{5.4}$$

where $J_c(x_c)$ is the field distribution across the source of the illuminating wave. Substituting (5.1), (5.2), and (5.4) into (5.3) leads to an expression for the transmitted wave

$$\psi(\overset{\cdot}{x}) = \frac{1}{\lambda f}\left(\frac{-i}{\lambda f}\right)^{\frac{1}{2}} \iiint_{-\infty}^{\infty} J_c(x_c)J_R^*(x_R)F(x_o)e^{-i(k/f)x(x-x_R+x_o)}\,dx_c\,dx_R\,dx_o.$$
(5.5)

To form an image with this transmitted wave we use a lens of focal length f as shown in Fig. 5.2. The resulting image $G(x_i)$ is formed in the plane $z = 2f$ and is given by

$$G(x_i) = \left(\frac{-i}{\lambda f}\right)^{\frac{1}{2}} \int_{-\infty}^{\infty} \psi(x)e^{-i(k/f)xx_i}\,dx.$$
(5.6)

We have written the limits on this integral as $\pm\infty$ even though the range of x will be limited by the finite dimensions of the hologram. The infinite limits will represent a good approximation as long as the hologram size is large enough so that the effects of diffraction are negligible compared to the effects caused by the extent of the sources J_R and J_c. We will compute these effects separately, beginning with those due to the finite size of the sources in the absence of diffraction. Later we will return to (5.6) and determine the effects of diffraction.

We can best determine these effects by finding the *line spread function* for the holographic system. This is the irradiance distribution across the image of an infinitely thin line object. This will give us the best estimate of the resolution capabilities of the system. We first find the amplitude distribution in the image of a line. Substituting (5.5) into (5.6) gives

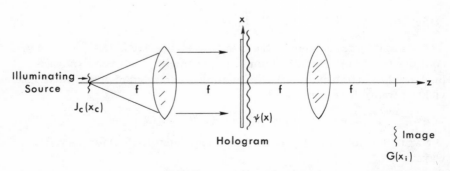

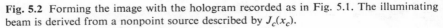

Fig. 5.2 Forming the image with the hologram recorded as in Fig. 5.1. The illuminating beam is derived from a nonpoint source described by $J_c(x_c)$.

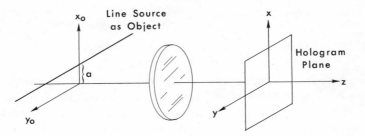

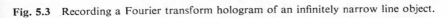

Fig. 5.3   Recording a Fourier transform hologram of an infinitely narrow line object.

$$G(x_i) = \frac{-1}{\lambda^2 f^2} \iiiint_{-\infty}^{\infty} J_c(x_c) J_R^*(x_R) F(x_o) e^{-i(k/f)x(x_c - x_R + x_o + x_i)} \, dx_c \, dx_R \, dx_o \, dx.$$

$$(5.7)$$

Writing a line object located a distance a above the axis (see Fig. 5.3) as

$$F(x_o) = \delta(x_o - a), \qquad (5.8)$$

where $\delta(x)$ is the Dirac delta function, the amplitude distribution in the image becomes

$$G_l(x_i)$$

$$= \frac{-1}{\lambda^2 f^2} \iiiint_{-\infty}^{\infty} J_c(x_c) J_R^*(x_R) \delta(x_o - a) e^{-i(k/f)x(x_c - x_R + x_o + x_i)} dx_c \, dx_R \, dx_o \, dx$$

$$= \frac{-1}{\lambda^2 f^2} \iiint_{-\infty}^{\infty} J_c(x_c) J_R^*(x_R) e^{-i(k/f)x(x_c - x_R + a + xi)} dx_c \, dx_R \, dx, \qquad (5.9)$$

where we have written $G_l(x_i)$ to denote that this is the image of a line object. Using the relation

$$\delta(x) = \frac{1}{2\pi} \int_{-\infty}^{\infty} e^{-ikx} \, dk, \qquad (5.10)$$

we see that

$$\int_{-\infty}^{\infty} e^{-i(k/f)x(x_c - x_R + a + xi)} \, dx = 2\pi \delta \left[\frac{k}{f} (x_c - x_R + a + x_i) \right]$$

$$= \lambda f \delta(x_c - x_R + a + x_i) \qquad (5.11)$$

so that

$$G_l(x_i) = \frac{-1}{\lambda f} \iint_{-\infty}^{\infty} J_c(x_c) J_R^*(x_R) \delta(x_c - x_R + a + x_i) \, dx_c \, dx_R. \quad (5.12)$$

Integrating this over x_c gives

$$G_I(x_i) = \frac{-1}{\lambda f} \int_{-\infty}^{\infty} J_c(x_R - x_i - a) J_R^*(x_R)\, dx_R. \tag{5.13}$$

Hence we see that the amplitude spread function of the holographic system is given by a correlation of the reference and illuminating beam source distributions, J_c and J_R. The line spread function is $|G_I(x_i)|^2$. A full three-dimensional analysis would give rise to a two-dimensional correlation integral in place of (5.13) and would lead to the point spread function, or impulse response, of the holographic system. The correlation integral indicates that, in general, if the reference source has a width A and the illuminating source a width B, then the line spread function (in amplitude) of the system will have a width of the order of $A + B$—the maximum possible resolution of this system will be approximately $1/A + B$ resolvable lines per unit distance.

If the illuminating beam is derived from a point source (or in reality a very small one), but the reference source has a finite extent, then

$$J_c(x_c) = \delta(x_c) \tag{5.14}$$

and

$$G_I(x_i) = \frac{-1}{\lambda f} \int_{-\infty}^{\infty} \delta(x_R - x_i - a) J_R^*(x_R)\, dx_R$$

$$= \frac{-1}{\lambda f} J_R^*(x_i + a) \tag{5.15}$$

so that an image of a line at a distance a above the axis is the conjugate of the reference source distribution centered a distance a below the axis. Hence we can say that for illumination of the hologram with an ideal plane or spherical wave, the minimum resolvable line image can be no less than the extent of the reference source. Conversely, we can also say that if the reference wave is ideally plane or spherical, but the illuminating wave is derived from other than a point source, then the line object will image as the illuminating source distribution $J_c(x_c)$.

We next wish to determine the line spread function for diffraction-limited imagery. We begin by noting that in the ideal case both J_c and J_R are so small that we can write

$$J_R(x_R) = \delta(x_R)$$

and
$$\tag{5.16}$$

$$J_c(x_c) = \delta(x_c).$$

With this approximation (5.7) becomes

$$G(x_i) = -\frac{1}{\lambda^2 f^2} \iiiint_{-\infty}^{\infty} \delta(x_c)\,\delta(x_R)F(x_o)e^{-i(k/f)x(x_c-x_R+x_o+x_i)}dx_c\,dx_R\,dx_o\,dx$$

$$= -\frac{1}{\lambda^2 f^2} \iint_{-\infty}^{\infty} F(x_o)e^{-i(k/f)x(x_o+x_i)}\,dx_o\,dx. \tag{5.17}$$

As long as the limits on the x-integration are $\pm\infty$, we may use the definition (5.10) and write

$$G(x_i) = -\frac{2\pi}{\lambda^2 f^2} \int_{-\infty}^{\infty} F(x_o)\,\delta\left[\frac{k}{f}(x_o+x_i)\right]dx_o$$

$$= -\frac{2\pi}{\lambda^2 f^2}\frac{f}{k} \int_{-\infty}^{\infty} F(x_o)\,\delta(x_o+x_i)\,dx_o$$

$$= -\frac{1}{\lambda f}F(-x_i) \tag{5.18}$$

and the imagery is perfect. To account for the effects of diffraction, we note that in any real case the hologram plane only extends from $-H$ to H so that (5.17) should be written

$$G(x_i) = \frac{-1}{\lambda^2 f^2} \int_{-\infty}^{\infty} F(x_o) \left\{ \int_{-H}^{H} e^{-i(k/f)x(x_o+x_i)}\,dx \right\} dx_o$$

$$= \frac{-2H}{\lambda^2 f^2} \int_{-\infty}^{\infty} F(x_o)\operatorname{sinc}\left[\frac{kH}{f}(x_o+x_i)\right]dx_o, \tag{5.19}$$

where $\operatorname{sinc} x = \sin x/x$.

For an infinitely narrow line object located at $(x_o, z_o) = (a, -2f)$, we have

$$F(x_o) = \delta(x_o - a) \tag{5.20}$$

and the diffraction-limited, line amplitude spread function becomes

$$G_l(x_i) = \frac{-2H}{\lambda^2 f^2} \int_{-\infty}^{\infty} \delta(x_o - a)\operatorname{sinc}\left[\frac{kH}{f}(x_o+x_i)\right]dx_o$$

$$= \frac{-2H}{\lambda^2 f^2}\operatorname{sinc}\left[\frac{kH}{f}(x_i+a)\right]. \tag{5.21}$$

The line spread function of the system is now just

$$|G_l(x_i)|^2 = \frac{4H^2}{\lambda^4 f^4}\operatorname{sinc}^2\left[\frac{kH}{f}(x_i+a)\right]. \tag{5.22}$$

If a three-dimensional analysis had been made, the image of a point object located at $x_o = a$, $y_o = b$ would be given by

$$G_p(x_i, y_i) = \frac{-4H^2}{\lambda^4 f^4} \operatorname{sinc}\left[\frac{kH}{f}(x_i + a)\right] \operatorname{sinc}\left[\frac{kH}{f}(y_i + b)\right] \quad (5.23)$$

and the point spread function for the system would be $|G_p(x_i, y_i)|^2$. The line spread function (5.22) is centered about the point $x_i = -a$. The width of the line image can be taken as the distance between the zeros on either side of the central maximum of the sinc function. These occur for the argument equal to $\pm\pi$:

$$\frac{kH}{f}(x_i + a) = \pm\pi. \quad (5.24)$$

The width of the line image is thus $f\lambda/H$. Hence the number of resolvable lines per unit distance will be of the order of $H/f\lambda$.

In the most common situation both the reference and illuminating beams are derived from the same source, so that $J_c = J_R$. In this case (5.13) becomes

$$G_l(x_i) = \frac{-1}{\lambda f} \int_{-\infty}^{\infty} J_R(x_R - x_i - a) J_R^*(x_R)\, dx_R. \quad (5.25)$$

Some interesting conclusions may be drawn from this relationship, but first we write (5.2) as a Fourier transform. Since (5.2) is in the same form as (A.18) of the Appendix (except for the constant phase factor e^{2ikf}), we can see that the desired form is

$$V(\nu) = \int_{-\infty}^{\infty} J_R(x_R) e^{-2\pi i \nu x_R}\, dx_R \quad (5.26)$$

where

$$V(\nu) = \left(\frac{-i}{\lambda f}\right)^{-\frac{1}{2}} \cdot R(\lambda f \nu) \quad (5.27)$$

and

$$\nu = \frac{x}{\lambda f}. \quad (5.28)$$

By Fourier inversion of (5.26), we can write

$$J_R(x_R) = \int_{-\infty}^{\infty} V(\nu) e^{2\pi i \nu x_R}\, d\nu \quad (5.29)$$

so that from (5.25)

$$-\lambda f G_l(x_i) = \iint_{-\infty}^{\infty} V(\nu) e^{2\pi i \nu(x_R - x_i - a)}\, d\nu\, dx_R \int_{-\infty}^{\infty} V^*(\nu') e^{-2\pi i \nu' x_R}\, d\nu'. \quad (5.30)$$

Interchanging the order of integration and combining the exponentials, we get

$$-\lambda f G_l(x_i) = \iiint_{-\infty}^{\infty} V(\nu)V^*(\nu')e^{-2\pi i\nu(x_i+a)}e^{-2\pi ix_R(\nu'-\nu)}\, d\nu\, d\nu'\, dx_R$$

$$= \iint_{-\infty}^{\infty} V(\nu)V^*(\nu')e^{-2\pi i\nu(x_i+a)}\delta(\nu'-\nu)\, d\nu\, d\nu'$$

$$= \int_{-\infty}^{\infty} |V(\nu)|^2 e^{-2\pi i\nu(x_i+a)}d\nu. \tag{5.31}$$

Recall that $G_l(x_i)$ is the line amplitude spread function of the holographic system and that $V(\nu)$ is closely related to the light distribution at the hologram plane caused by the reference beam alone; $V(\nu)$ is the Fourier transform of $J_R(x_R)$ which is the field distribution across the reference source. Since G_l is essentially an autocorrelation function (5.25), it is not surprising that it can be written as the Fourier transform of the energy spectrum $|V(\nu)|^2$. It is interesting to note the form of the line amplitude spread function when $V(\nu)$ is band limited. Suppose that the hologram extends from $-H$ to H so that the range of spatial frequencies that can be recorded is only $-\nu_m$ to ν_m where

$$\nu_m = \frac{H}{\lambda f}. \tag{5.32}$$

Suppose that the reference source is a large diffuser, large enough that the complete frequency range $2\nu_m$ is covered (Fig. 5.4). If the diffuser is perfect, that is, completely random, $J_R(x_R)$ is in the form of a white noise so that

$$|V(\nu)|^2 = V_0, \qquad -\nu_m \le \nu \le \nu_m$$

$$= 0 \qquad \text{elsewhere.} \tag{5.33}$$

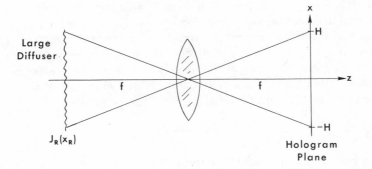

Fig. 5.4   The use of a large diffuser as the reference source.

In this case (5.31) becomes

$$-\lambda f G_l(x_i) = \int_{-\nu_m}^{\nu_m} V_0 e^{-2\pi i \nu(x_i + a)} \, d\nu \tag{5.34}$$

so that

$$G_l(x_i) = -V_0 \frac{2\nu_m}{\lambda f} \operatorname{sinc} [2\pi \nu_m(x_i + a)]. \tag{5.35}$$

Using (5.32) we find

$$G_l(x_i) = -V_0 \frac{2H}{\lambda^2 f^2} \operatorname{sinc} \left[\frac{kH}{f}(x_i + a) \right] \tag{5.36}$$

which, except for the constant V_0, is exactly the same as (5.21). This means that the image resolution should be the same whether a point reference source or a perfectly diffuse one is used. The assumption that the diffuser is perfect is impossible, however, since this would require that both the diffuser and lens of Fig. 5.4 be infinitely large. Any finite-sized diffuser leads to a speckle pattern in the hologram plane which means that $|V(\nu)|^2$ is not constant. This speckle is, of course, just the random interference pattern that results from the interference of all pairs of points of the diffuser. In actual practice, however, the resolution that can be achieved with a diffuse reference beam is about equal to that which can be obtained with a small reference source. The point spread function for such a system is essentially the diffraction-limited point spread function, surrounded by uniform flare.

There is an important reciprocal relationship between source size and image resolution which may be stated simply as follows. If the correlation function (5.25) is very sharply peaked, indicating either a very small or very large and diffuse reference source distribution, the size of the hologram that can be filled with light is very large, resulting in only a small amount of light spread caused by diffraction. Hence the resolution will be high. On the other hand, if the correlation function (5.25) is very broad, indicating a large source, only a small hologram area can be filled, resulting in poor resolution because of diffraction.

As an example of this effect, consider the common situation where a microscope objective is used to diverge the reference beam so that it will fill the hologram. The spot at the focus of the objective then serves as the source for the reference beam. For the TEM$_{oo}$ laser mode, the spot size at the focus of the objective is given by

$$w_0 = \frac{\lambda}{\pi \, \text{N.A.}} \tag{5.37}$$

where N.A. is the numerical aperture of the objective [see (6.28) et seq.] and w_0 is the radius of the spot measured to the $1/e$ amplitude points.

But w_0 is related to the hologram size $2H$ and the focal length f of the lens in Fig. 5.1 through

$$\text{N.A.} = \frac{H}{f} = \frac{\lambda}{\pi w_0} \tag{5.38}$$

so that

$$w_0 = \frac{\lambda f}{\pi H}. \tag{5.39}$$

The width of the correlation peak of (5.25) is thus roughly

$$4w_0 = \frac{4\lambda f}{\pi H}. \tag{5.40}$$

The half width of a line image for a diffraction-limited hologram is $f\lambda/H$ (see Eq. 5.24). Comparison of this result with (5.40) indicates that the two resolution limits are essentially equal. Therefore nothing will be gained in terms of image resolution by using a higher N.A. objective than is necessary just to fill the hologram with light. Overfilling the hologram by use of a high N.A. objective will not increase the resolution, even though the reference source size will be small.

5.2   REFERENCE AND ILLUMINATING
BEAM BANDWIDTH

It is well known that high-quality holographic images require a narrow-band light source, both for recording and illuminating the hologram. An ideal light source will be a point source of zero size and zero bandwidth. Actual sources used in practice are only approximations of this ideal. Sharpness in an image is always limited by diffraction, so the best possible light source need be no smaller than the size dictated by diffraction effects, as previously discussed. Further, the bandwidth need not be zero but only small enough so that any spreading of light in the image caused by finite bandwidth will be small compared to the spreading due to diffraction.

We can obtain an estimate of the resolution-bandwidth relationship by considering a simple diffraction grating model of a hologram. Suppose two plane waves are incident on the hologram plane at some angle α_R as shown in Fig. 5.5. The resulting interferogram constitutes a Fourier transform hologram of a point object, and the fringes recorded on the hologram vary spatially in transmittance in an approximately sinusoidal manner. The spatial frequency of these fringes is given by

$$\nu_s = \frac{\sin \alpha_R}{\lambda} \tag{5.41}$$

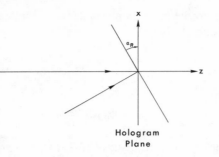

Fig. 5.5   A simple plane wave hologram.

where λ is the recording wavelength. If the recording light has a finite bandwidth $\Delta\lambda_R$, then

$$\lambda_R - \frac{\Delta\lambda_R}{2} \leq \lambda \leq \lambda_R + \frac{\Delta\lambda_R}{2} \qquad (5.42)$$

where λ_R is the central wavelength of the recording light. This wavelength spread gives rise to a spread in recorded spatial frequency given by

$$\Delta\nu_s = -\frac{\sin \alpha_R}{\lambda_R{}^2} \Delta\lambda_R. \qquad (5.43)$$

To reconstruct, we illuminate the hologram with a plane wave incident at the angle $\alpha_c = \alpha_R$ and wavelength λ (Fig. 5.6) such that

$$\lambda_c - \frac{\Delta\lambda_c}{2} \leq \lambda \leq \lambda_c + \frac{\Delta\lambda_c}{2} \qquad (5.44)$$

where $\Delta\lambda_c$ is the bandwidth of the illuminating beam and λ_c the central

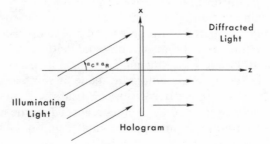

Fig. 5.6 Diffraction of the primary wave from the grating (hologram) recorded as in Fig. 5.5. The diffracted wave makes an angle $\alpha_i = 0$ with the z-axis if the reference and illuminating beams are monochromatic.

wavelength. The angle of diffraction α_i of the first order (primary wave) is given by

$$\sin \alpha_i = \lambda \nu_s - \sin \alpha_c \qquad (5.45)$$

which is equal to zero for monochromatic light, as shown in Fig. 5.6. Due to the spread in wavelengths of the illuminating and recording beams, however, there will be a spread in diffraction angles

$$\Delta(\sin \alpha_i) = (\sin \alpha_i)_{max} - (\sin \alpha_i)_{min}, \qquad (5.46)$$

where

$$(\sin \alpha_i)_{max} = \lambda_{max}\nu_{s(max)} = \left(\lambda_c + \frac{\Delta\lambda_c}{2}\right)\left(\frac{\sin \alpha_R}{\lambda_R} + \frac{\sin \alpha_R}{2\lambda_R{}^2}\Delta\lambda_R\right) \qquad (5.47)$$

and

$$(\sin \alpha_i)_{min} = \lambda_{min}\nu_{s(min)} = \left(\lambda_c - \frac{\Delta\lambda_c}{2}\right)\left(\frac{\sin \alpha_R}{\lambda_R} - \frac{\sin \alpha_R}{2\lambda_R{}^2}\Delta\lambda_R\right) \qquad (5.48)$$

and therefore

$$\Delta(\sin \alpha_i) = \frac{\lambda_c}{\lambda_R}\left(\frac{\Delta\lambda_R}{\lambda_R} + \frac{\Delta\lambda_c}{\lambda_c}\right)\sin \alpha_R. \qquad (5.49)$$

We see that the finite bandwidths of both recording and illuminating beams have led to an angular spread in the reconstructed object wave. In terms of the optical frequencies involved, this spread is given by

$$\Delta(\sin \alpha_i) = \frac{\nu_R}{\nu_c}\left(\frac{\Delta\nu_c}{\nu_c} + \frac{\Delta\nu_R}{\nu_R}\right)\sin \alpha_R. \qquad (5.50)$$

Often the reference and illuminating beams are identical. In this case $\nu_R = \nu_c$ and $\Delta\nu_R = \Delta\nu_c$, so

$$\Delta(\sin \alpha_i) = 2\frac{\Delta\nu_R}{\nu_R}\sin \alpha_R. \qquad (5.51)$$

For a multimode He-Ne gas laser the effective $\Delta\nu_R$ is of the order of 1.5×10^9 Hz. This leads to

$$\Delta(\sin \alpha_i) \approx 6 \times 10^{-6} \sin \alpha_R \approx \Delta\alpha_i. \qquad (5.52)$$

This is very small, especially since the angular spread due to diffraction, $\Delta\theta_D$, is given by

$$\Delta\theta_D \approx \frac{\lambda}{2H} = \frac{c}{2H\nu_c} \qquad (5.53)$$

where $2H$ is the hologram aperture. For the He-Ne gas laser the angular spread due to finite bandwidth becomes equal to the spread caused by diffraction for a hologram aperture

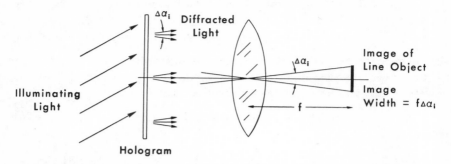

Fig. 5.7 The line spread function for a hologram system in which the hologram is recorded and illuminated with light of nonzero bandwidth. This gives rise to a spread in diffraction angles $\Delta\alpha_i$ and a line spread function of width $f\Delta\alpha_i$.

$$2H \cong 20 \text{ cm.} \tag{5.54}$$

For the Argon laser the situation is slightly worse because of a factor of about three in $\Delta\nu_R$. For this laser the two angular spreads are equal for an aperture of

$$2H \cong 5 \text{ cm.} \tag{5.55}$$

The spread of directions $\Delta\alpha_i$ of the diffracted wave means that this wave will focus to a line width $f\Delta\alpha_i$ (Fig. 5.7). This will be the width of the line spread function of the system.

Although we have confined the discussion to a simple diffraction grating hologram, the results are easily extended to more complex object distributions. For a Fourier transform hologram of an extended object, each object point records as a single spatial frequency at the hologram plane, so that the foregoing analysis can be applied directly. For a Fresnel hologram we may use these results if we consider the hologram a superposition of very many simple gratings. Each object point is now recorded as a range of spatial frequencies at the hologram plane. The range of spatial frequencies given by 5.43 can now be regarded as the spread in recorded frequency at each small element of area of the hologram. There will now be a different $\Delta\alpha_i$ at each point of the hologram, but the resulting spread function will be substantially the same as for a Fourier hologram for most recording configurations. The line spread function for a Fresnel hologram will have an approximate width given by $\overline{\Delta\alpha_i} \cdot z_o$, where $\overline{\Delta\alpha_i}$ is the average spread in the diffracted beam and z_o is the image distance.

5.3   EFFECT OF THE RECORDING MEDIUM [1, 2]

5.3.1   Introduction

We have been tacitly assuming that the recording medium is capable of resolving all of the spatial frequencies of interest, except possibly for a cutoff due to the finite extent of the hologram. This assumption, of course, represents an idealized situation, since every recording medium will have some upper limit on the spatial frequencies it is capable of recording. One of the most useful measures of the ability of a recording medium to resolve fine detail is its modulation transfer function, or in short, its MTF. This is just the ratio of output to input sine wave modulations, expressed as a function of spatial frequency. If it is defined in one dimension, it is just the Fourier transform of the line spread function. The purpose of this section, then, is to determine the effect of the MTF of the recording medium on the resolution in the holographic image.

5.3.2   Fourier Transform Holograms

Again we consider a basic Fourier transform arrangement such as shown in Fig. 5.8. The reference beam makes an angle α_R with the axis as shown; the complex object distribution is denoted by $F(x_o)$. The object is situated at $z_o = -2f$, a distance f from a well-corrected lens of focal length f. The hologram plane is a distance f from the lens, so that we may write for the total disturbance at the hologram plane

$$H(x) = R(x) + O(x) = e^{-ikx\sin\alpha_R} + \left(\frac{-i}{\lambda f}\right)^{\frac{1}{2}} \int_{-\infty}^{\infty} F(x_o)e^{-i(k/f)xx_o}dx_o. \quad (5.56)$$

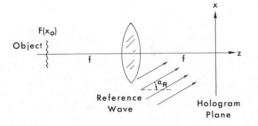

Fig. 5.8   A basic Fourier transform hologram arrangement.

The incident exposure will be proportional to $|H(x)|^2$, and the effective exposure, due to the limited MTF of the recording medium, will be a convolution of the incident exposure and the spread function of the recording medium. In one dimension, then, the effective exposure is

$$E(x) = \int_{-\infty}^{\infty} S(u) |H(x-a)|^2 \, du, \tag{5.57}$$

where $S(u)$ is the line spread function of the medium.* In order to eliminate as many unimportant constants as possible, we will assume that, after processing, the resulting amplitude transmittance of the hologram is simply equal to the exposure. In this case

$$t(x) = E(x) = \int_{-\infty}^{\infty} S(u)[|R(x-u)|^2 + |O(x-u)|^2$$

$$+ R^*(x-u)O(x-u) + R(x-u)O^*(x-u)] \, du$$

$$\equiv I_0 + I_1 + I_2 \tag{5.58}$$

where

$$I_0 = \int_{-\infty}^{\infty} S(u)[|R(x-u_J)|^2 + |O(x-u)|^2] \, du \tag{5.59}$$

$$I_1 = \int_{-\infty}^{\infty} S(u)[R^*(x-u)O(x-u)] \, du \tag{5.60}$$

$$I_2 = \int_{-\infty}^{\infty} S(u)_{\mathsf{L}} R(x-u)O^*(x-u)] \, du. \tag{5.61}$$

Identification of these three components follows easily from earlier discussion: I_0 is the zero-order bias plus flare light, I_1 is the primary image wave, and I_2 is the conjugate image wave. Considering only $I_1(x)$ for the primary image, we have

* There is some question as to whether this convolution can be carried out as indicated or whether one must use an *amplitude spread function* convolved with the light *amplitude* at the hologram. The reason for the question is that the exposing light is both spatially and temporally coherent, which means that one must add amplitudes and then square the sum to arrive at the irradiance. On the other hand, (5.57) implies that each elementary line in the object images to an irradiance distribution and that the elementary $S(u)$ functions simply add to give the resultant irradiance distribution. Whether or not one is justified in using (5.57) will have to await future developments; the data taken by van Ligten [1] show a discrepancy with the theory which might be explained along these lines, although Vander Lugt's [3] work shows good agreement with theory.

$$I_1(x) = \int_{-\infty}^{\infty} S(u)[R^*(x-u)O(x-u)] \, du$$

$$= \left(\frac{-i}{\lambda f}\right)^{\frac{1}{2}} \int_{-\infty}^{\infty} S(u)e^{iks\sin\alpha_R(x-u)} \int_{-\infty}^{\infty} F(x_o)e^{-i(k/f)x_o(x-u)} \, dx_o \, du \qquad (5.62)$$

$$= \left(\frac{-i}{\lambda f}\right)^{\frac{1}{2}} e^{ikx\sin\alpha_R} \int_{-\infty}^{\infty} S(u)e^{i(k/f)u(x_o-f\sin\alpha_R)} \, du \int_{-\infty}^{\infty} F(x_o)e^{-i(k/f)xx_o} \, dx_o.$$

Since the MTF of the recording medium is just the Fourier transform of the line spread function $S(u)$, we have

$$\text{MTF} = \tilde{S}(\nu) = \int_{-\infty}^{\infty} S(u)e^{-2\pi i\nu u} \, du, \qquad (5.63)$$

where the $\sim$ implies a Fourier transform. Thus we have

$$\tilde{S}\left[\frac{k}{2\pi f}(f\sin\alpha_R - x_o)\right] = \int_{-\infty}^{\infty} S(u)e^{i(k/f)u(x_o-f\sin\alpha_R)} \, du \qquad (5.64)$$

and (5.62) becomes

$$I_1(x) = \left(\frac{-i}{\lambda f}\right)^{\frac{1}{2}} e^{-ikx\sin\alpha_R} \int_{-\infty}^{\infty} \tilde{S}\left[\frac{k}{2\pi f}(f\sin\alpha_R - x_o)\right] F(x_o)e^{-i(k/f)xx_o} \, dx_o.$$

$$(5.65)$$

The image is formed in the usual way—a second lens takes the Fourier transform of the transmitted field distribution in the x plane. If the hologram is illuminated with a wave identical to the reference wave, then the transmitted wave is

$$\psi(x) = e^{-ikx\sin\alpha_R}I_1(x) \qquad (5.66)$$

and the image is

$$G(x_i) = \left(\frac{-i}{\lambda f}\right)^{\frac{1}{2}} \int_{-\infty}^{\infty} \psi(x)e^{-i(k/f)xx_i} \, dx. \qquad (5.67)$$

Substituting (5.65) into (5.66), we obtain

$$G(x_i) = \frac{-i}{\lambda f} \iint_{-\infty}^{\infty} \tilde{S}\left[\frac{k}{2\pi f}(f\sin\alpha_R - x_o)\right] F(x_o)e^{-i(k/f)x(x_i+x_o)} \, dx \, dx_o. \qquad (5.68)$$

Using the relation

$$\delta(x) = \frac{1}{2\pi} \int_{-\infty}^{\infty} e^{-ikx} \, dk \qquad (5.69)$$

for the Dirac δ-function, the image becomes

$$G(x_i) = \frac{-2\pi i}{\lambda f} \int_{-\infty}^{\infty} \tilde{S}\left[\frac{k}{2\pi f}(f \sin \alpha_R - x_o)\right] F(x_o)\, \delta\left[\frac{k}{f}(x_o + x_i)\right] dx_o$$

$$= -i\tilde{S}\left[\frac{\sin \alpha_R}{\lambda} + \frac{x_i}{\lambda f}\right] F(-x_i). \tag{5.70}$$

Thus we find that for a Fourier transform hologram, recorded on a medium with limited MTF, the image resolution is not affected, but the field of view is limited.

As a simple example of this effect, consider an MTF given by

$$\tilde{S}(\nu) = 1, \qquad -\nu_c \leq \nu \leq \nu_c$$

$$= 0 \qquad \text{otherwise.} \tag{5.71}$$

We wish to make a hologram of a sinusoidly varying object*

$$F(x_o) = 1 + M \cos 2\pi \nu_o x_o \tag{5.72}$$

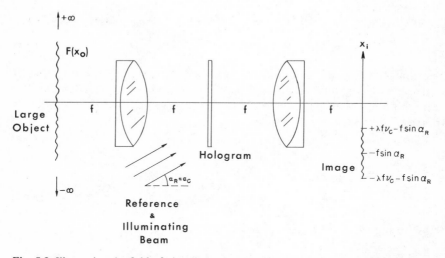

Fig. 5.9 Illustrating the field-of-view limitation caused by recording a hologram of a large object on a recording medium which can only record spatial frequencies up to ν_c. Although the object extends between $\pm \infty$, the image in the x_i plane extends only from $\lambda f \nu_c - f \sin \alpha_R$ to $-\lambda f \nu_c - f \sin \alpha_R$.

* Strictly speaking, there should be a random phase factor multiplying this expression, implying diffuse illumination, because if this object were specularly illuminated, $O(x)$ would consist only of three very small points of light and no hologram could be formed.

where M is the modulation of the pattern and ν_o the spatial frequency. In (5.70) we see that the image will be identical to the object in the range

$$-\lambda f \nu_c - f \sin \alpha_R \leq x_i \leq \lambda f \nu_c - f \sin \alpha_R. \tag{5.73}$$

We see that the image is independent of the object frequency ν_o but that it is limited in extent in an asymmetrical manner (see Fig. 5.9). The object is assumed to extend to $\pm \infty$ in the x_o plane, but only a finite portion is imaged in the x_i plane. The diameter of both lenses and the hologram are assumed to be large enough so that aperture effects can be neglected.

5.3.3 Fresnel Holograms

Consider the arrangement illustrated in Fig. 5.10. A plane object is defined by the complex amplitude distribution $F(x_o)$ and is located in the x_o plane a distance z_o from the hologram. A plane reference wave is used, and this wave makes an angle α_R with the axis. We assume that z_o is large enough that we may use the Fresnel approximations and write, for the amplitude distribution at the hologram due to the object,

$$O(x) = \left(\frac{-i}{\lambda}\right)^{\frac{1}{2}} \int_{-\infty}^{\infty} F(x_o) \frac{e^{ikr}}{r} dx_o. \tag{5.74}$$

The distance between a point x on the hologram and a point x_o on the object is denoted by r and can be written as

$$r = \pm[(x - x_o)^2 + z_o^2]^{\frac{1}{2}} \approx \pm z_o \pm \frac{(x - x_o)^2}{2z_o}. \tag{5.75}$$

Since $|z_o| = -z_o$ for our coordinate system, we must choose the negative signs to retain our phase convention [c.f. (A.5)].

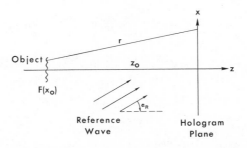

Fig. 5.10 The notation used to describe the recording of a Fresnel hologram.

The reference w: written as usual

$$R(x) = e^{-ikx\sin\alpha_R} \tag{5.76}$$

and we again consider only the primary image term

$$I_1(x) = \int_{-\infty}^{\infty} S(u)[R^*(x - u)O(x - u)]\, du.$$

Substituting (5.74), (5.75), and (5.76), this becomes

$$I_1(x) = \left(\frac{i}{\lambda z_o}\right)^{\frac{1}{2}} \int_{-\infty}^{\infty} S(u)e^{ik(x-u)\sin\alpha_R}$$

$$\times \left[\int_{-\infty}^{\infty} F(x_o)e^{-ikz_o}e^{-ik(x-u-x_o)^2/2z_o}\, dx_o\right] du, \tag{5.77}$$

where we have written $1/r \approx 1/-z_o$. Now define

$$\tilde{F}(\nu) = \int_{-\infty}^{\infty} F(x_o)e^{-2\pi i\nu x_o}\, dx_o \tag{5.78}$$

so that

$$F(x_o) = \int_{-\infty}^{\infty} \tilde{F}(\nu)e^{2\pi i\nu x_o}\, d\nu. \tag{5.79}$$

We then have

$$I_1(x) = K\int_{-\infty}^{\infty} S(u)e^{ik(x-u)\sin\alpha_R} \int\int_{-\infty}^{\infty} \tilde{F}(\nu)e^{2\pi i\nu x_o}e^{-ik(x-u-x_o)^2/2z_o}\, d\nu\, dx_o\, du, \tag{5.80}$$

where we have defined

$$K = \left(\frac{i}{\lambda z_o}\right)^{\frac{1}{2}}e^{-ikz_o}. \tag{5.81}$$

Now

$$\int_{-\infty}^{\infty} \exp\left[-i\frac{k}{2z_o}(x - u - x_o)^2\right]e^{2\pi i\nu x_o}\, dx_o = \left(\frac{\lambda z_o}{i}\right)^{\frac{1}{2}}e^{2\pi i\nu(x-u)}e^{2i(z_o/k)\pi^2\nu} \tag{5.82}$$

so

$$I_1(x) = e^{-ikz_o}\int_{-\infty}^{\infty} S(u)e^{ik(x-u)\sin\alpha_R}\left[\int_{-\infty}^{\infty} \tilde{F}(\nu)e^{2\pi i\nu(x-u)}e^{2i(z_o/k)\pi^2\nu^2}\, d\nu\right] du$$

$$= e^{-ikz_o}e^{ikx\sin\alpha_R}\int_{-\infty}^{\infty} S(u)\exp\left[-iku\left(\sin\alpha_R + \frac{2\pi\nu}{k}\right)\right] du$$

$$\times \int_{-\infty}^{\infty} \tilde{F}(\nu)e^{2\pi i\nu x}e^{2i(z_o/k)\pi^2\nu^2}\, d\nu. \tag{5.83}$$

Now define

$$\tilde{S}(\nu) = \int_{-\infty}^{\infty} S(u)e^{-2\pi i\nu u}\, du \equiv \text{MTF} \qquad (5.84)$$

so that

$$\tilde{S}\left[\nu + \frac{k\sin\alpha_R}{2\pi}\right] = \int_{-\infty}^{\infty} S(u)e^{-iku[\sin\alpha_R+(2\pi\nu/k)]}\, du \qquad (5.85)$$

and

$$I_1(x) = e^{-ikz_0}e^{ikx\sin\alpha_R}\int_{-\infty}^{\infty}\tilde{S}\left(\nu + \frac{\sin\alpha_R}{\lambda}\right)\tilde{F}(\nu)e^{2\pi i\nu x}e^{2i(z_0/k)\pi^2\nu^2}\, d\nu.$$

$$(5.86)$$

To reconstruct, we illuminate the hologram with a wave described by $C(x)$ in the hologram plane. The transmitted field distribution is then

$$\psi(x) = C(x)I_1(x), \qquad (5.87)$$

corresponding, of course, only to the primary image wave. The field in a plane x_i at a distance z_i from the hologram (Fig. 5.11) is given by

$$G(x_i) = \left(\frac{-i}{\lambda}\right)^{1/2}\int_{-\infty}^{\infty}\psi(x)\frac{e^{iks}}{s}\, dx \qquad (5.88)$$

where s is the distance between x and x_i:

$$s = \pm[(x - x_i)^2 + z_i^2]^{1/2} \approx \pm z_i \pm \frac{(x - x_i)^2}{2z_i}, \qquad (5.89)$$

and this time we use the positive root.

If we now assume that the illuminating wave $C(x)$ is identical to the

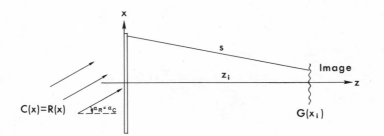

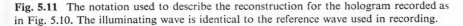

Fig. 5.11   The notation used to describe the reconstruction for the hologram recorded as in Fig. 5.10. The illuminating wave is identical to the reference wave used in recording.

reference wave (5.76), $C(x) = R(x)$ and substitution of (5.87) into (5.88) yields, with the approximation (5.89),

$$G(x_i) = \left(\frac{-i}{\lambda z_i}\right)^{\frac{1}{2}} \int_{-\infty}^{\infty} R(x)I_1(x)e^{iks} \, dx$$

$$= \left(\frac{-i}{\lambda z_i}\right)^{\frac{1}{2}} e^{-ik(z_o - z_i)} \iint_{-\infty}^{\infty} \tilde{S}\left(\nu + \frac{\sin \alpha_R}{\lambda}\right)$$

$$\times \tilde{F}(\nu)e^{2\pi i \nu x} e^{2i(z_o/k)\pi^2 \nu^2} \exp\left[ik\frac{(x - x_i)^2}{2z_i}\right] d\nu \, dx \quad (5.90)$$

where we have used $1/s \approx 1/z_i$. But

$$\int_{-\infty}^{\infty} \exp\left[i\frac{k}{2z_i}(x - x_i)^2\right] e^{2\pi i \nu x} \, dx = (i\lambda z_i)^{\frac{1}{2}} e^{2\pi i \nu x_i} e^{-2i(z_i/k)\pi^2 \nu^2}, \quad (5.91)$$

so

$$G(x_i) = e^{-ik(z_o - z_i)} \int_{-\infty}^{\infty} \tilde{S}\left(\nu + \frac{\sin \alpha_R}{\lambda}\right) \tilde{F}(\nu)e^{2\pi i \nu x_i} \exp\left[2i\frac{(z_o - z_i)}{k}\pi^2 \nu^2\right] d\nu$$

$$(5.92)$$

Since we require that the imagery be perfect for $\tilde{S}(\nu) = $ constant, we can see that a virtual image exists in the plane $z_i = z_o$, because in this plane $G(x_i)$ will be just the Fourier transform of the Fourier transform of the object distribution, which is the object distribution itself. Hence, for $z_i = z_o$,

$$G(x_i) = \int_{-\infty}^{\infty} \tilde{S}\left(\nu + \frac{\sin \alpha_R}{\lambda}\right) \tilde{F}(\nu)e^{2\pi i \nu x_i} \, d\nu. \quad (5.93)$$

To see the effect of limited MTF consider again the idealized case

$$\tilde{S}(\nu) = 1, \qquad -\nu_c \leq \nu \leq \nu_c$$

$$= 0 \qquad \text{otherwise} \quad (5.94)$$

with a single frequency object

$$F(x_o) = 1 + M \cos(2\pi\nu_o x_o) \quad (5.95)$$

with Fourier transform

$$\tilde{F}(\nu) = \delta(\nu) + \frac{M}{2}\delta(\nu - \nu_o) + \frac{M}{2}\delta(\nu + \nu_o). \quad (5.96)$$

According to (5.93), the image is

$$G(x_i) = \int_{-\infty}^{\infty} \tilde{S}\left(\nu + \frac{\sin \alpha_R}{\lambda}\right) \tilde{F}(\nu) e^{2\pi i \nu x_i} \, d\nu$$

$$= \int_{-\nu_c - \sin \alpha_R/\lambda}^{\nu_c - \sin \alpha_R/\lambda} \left[\delta(\nu) + \frac{M}{2} \delta(\nu - \nu_0) + \frac{M}{2} \delta(\nu + \nu_0)\right] e^{2\pi i \nu x_i} \, d\nu$$

$$\tag{5.97}$$

$$= \frac{M}{2} e^{-2\pi i \nu_0 x_i} \qquad \frac{\sin \alpha_R}{\lambda} - \nu_0 \leq \nu_c \leq \frac{\sin \alpha_R}{\lambda}$$

$$= 1 + M \cos(2\pi \nu_0 x_i) \qquad \nu_c \geq \nu_0 + \frac{\sin \alpha_R}{\lambda}$$

$$= 0 \qquad \nu_c \leq \frac{\sin \alpha_R}{\lambda} - \nu_0.$$

These three situations are easily interpreted. In the first case there is a uniform irradiance in the image plane. This occurs when the recording medium has been capable of recording the low-frequency sideband—carrier minus signal—but has been incapable of recording the carrier, $\sin(\alpha_R/\lambda)$. In the second case the complete image is formed, since the medium has resolved the carrier plus both sidebands. In the third case there is no light in the image plane since the recording medium has not even recorded the lowest frequency present.

This result points up one of the basic differences between a Fresnel hologram and the Fourier transform hologram described previously. In the latter the resolution was not affected by the limited MTF of the recording medium—only the field of view. In the present case the resolution is affected. Had the object of (5.95) contained a continuum of spatial frequencies, only those up to $\nu_{o(\text{max})}$ would have appeared in the image, that is, been resolved, where $\nu_{o(\text{max})} = \nu_c - \sin \alpha_R/\lambda$.

The line amplitude spread function of the holographic system will just be the image of an infinitely narrow line object. In this case, $F(x_o) = \delta(x_o)$ and $\tilde{F}(\nu) = 1$ so that (5.93) gives (again omitting e^{-2ikz_0})

$$G_l(x_i) = \int_{-\infty}^{\infty} \tilde{S}\left(\nu + \frac{\sin \alpha_R}{\lambda}\right) e^{2\pi i \nu x_i} \, d\nu \tag{5.98}$$

and by Fourier inversion of (5.84), the line amplitude spread function becomes

$$G_l(x_i) = \exp\left[-2\pi i \frac{x_i \sin \alpha_R}{\lambda}\right] S(x_i) \tag{5.99}$$

that is, just the line amplitude spread function of the recording medium, with an additional phase factor.

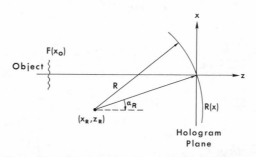

Fig. 5.12 Recording a Fresnel hologram with a spherical reference wave of radius R.

A still more general case which should be considered under the general topic of Fresnel holograms is that of a spherical reference wave, as indicated in Fig. 5.12. We assume that this wave has a radius R and originates from a point source located at (x_R, z_R). Its off-axis position is defined by an angle $\alpha_R = \sin^{-1}(x_R/R)$ so that, in one dimension,

$$R(x) = e^{-ik(x\sin\alpha_R + x^2/2R)}, \tag{5.100}$$

where we are assuming a uniform amplitude of unity across the hologram plane and have neglected terms of order $1/R^2$ in the exponential. For the primary image term we return to Eq. 5.60:

$$I_1(x) = \int_{-\infty}^{\infty} S(u)[R^*(x-u)O(x-u)]\, du.$$

Substituting (5.100) for $R(x)$ and writing

$$O(x) = \left(\frac{i}{\lambda z_o}\right)^{\frac{1}{2}} \int_{-\infty}^{\infty} F(x_o)e^{-ikz_o}\exp\left[-ik\frac{(x-x_o)^2}{2z_o}\right]dx_o \tag{5.101}$$

we obtain

$$I_1(x) = \left(\frac{i}{\lambda z_o}\right)^{\frac{1}{2}} e^{-ikz_o}\int_{-\infty}^{\infty} S(u)e^{ik(x-u)\sin\alpha_R}\exp\left[ik\frac{(x-u)^2}{2R}\right]$$
$$\times \int_{-\infty}^{\infty} F(x_o)\exp\left[-ik\frac{(x-u-x_o)^2}{2z_o}\right]dx_o\, du. \tag{5.102}$$

By interchanging the order of integration and regrouping some terms, (5.102) may be written

$$I_1(x) = \left(\frac{i}{\lambda z_o}\right)^{\frac{1}{2}} e^{-ikz_o} \int_{-\infty}^{\infty} F(x_o) e^{-ikx_o^2/2z_o}$$

$$\times \int_{-\infty}^{\infty} S(u) \exp\left[i\frac{k}{2}(x-u)^2\left(\frac{1}{R} - \frac{1}{z_o}\right)\right]$$

$$\times \exp\left[ik(x-u)\left(\frac{x_o}{z_o} + \sin \alpha_R\right)\right] du\, dx_o. \tag{5.103}$$

The u-integration is now in the form of a convolution of $S(u)$ and $\beta(u)$, where

$$\beta(u) = e^{ik(u^2/2R')} \exp\left[iku\left(\frac{x_o}{z_o} + \sin \alpha_R\right)\right], \tag{5.104}$$

with

$$\frac{1}{R'} = \frac{1}{R} - \frac{1}{z_o}. \tag{5.105}$$

This integration will be just the Fourier transform of the product of the Fourier transforms of $S(u)$ and $\beta(u)$ — $\tilde{S}(\nu)$ and $\tilde{\beta}(\nu)$, respectively. Now $\tilde{S}(\nu)$ is just the MTF of the recording medium and

$$\tilde{\beta}(\nu) = \int_{-\infty}^{\infty} \beta(u) e^{-2\pi i \nu u}\, du$$

$$= (i\lambda R')^{\frac{1}{2}} \exp\left[-i\left(\frac{kx_o}{z_o} + k\sin\alpha_R - 2\pi\nu\right)^2 \frac{R'}{2k}\right]. \tag{5.106}$$

Thus

$$I_1(x) = \left(\frac{i}{\lambda z_o}\right)^{\frac{1}{2}} e^{-ikz_o} \int_{-\infty}^{\infty} F(x_o) e^{-i(k/2z_o)x_o^2} \int_{-\infty}^{\infty} \tilde{S}(\nu)\tilde{\beta}(\nu) e^{2\pi i \nu x}\, d\nu\, dx_o$$

$$= i\left(\frac{R'}{z_o}\right)^{\frac{1}{2}} e^{-ikz_o} \int_{-\infty}^{\infty} F(x_o) e^{-i(k/2z_o)x_o^2}$$

$$\times \int_{-\infty}^{\infty} \tilde{S}(\nu) \exp\left[-i\left(\frac{kx_o}{z_o} + k\sin\alpha_R - 2\pi\nu\right)^2 \frac{R'}{2k}\right]$$

$$\times e^{2\pi i \nu x}\, d\nu\, dx_o. \tag{5.107}$$

By defining

$$\gamma \equiv k\sin\alpha_R - 2\pi\nu \tag{5.108}$$

and

$$\frac{1}{l} \equiv -\frac{1}{z_o} - \frac{R'}{2z_o^2}, \tag{5.109}$$

and doing some algebra, we find

$$I_1(x) = \left(\frac{R'}{z_o}\right)^{1/2} \frac{e^{ik(x\sin\alpha_R - z_o)}}{2\pi i}$$

$$\times \int_{-\infty}^{\infty} \tilde{S}\left(\frac{\sin\alpha_R}{\lambda} - \frac{\gamma}{2\pi}\right) \exp\left[-i\frac{R'}{2k}\left(1 + \frac{lR'}{z_o^2}\right)\gamma^2\right] e^{i\gamma x}$$

$$\times \int_{-\infty}^{\infty} F(x_o) \exp\left[i\frac{k}{2l}\left(\frac{lR'}{kz_o}\gamma - x_o\right)^2\right] d\gamma\, dx_o. \tag{5.110}$$

For reconstruction, let us assume that the illuminating wave is identical to the reference wave:

$$C(x) = R(x) = \exp\left[-ik\left(x\sin\alpha_R + \frac{x^2}{2R}\right)\right]. \tag{5.111}$$

The analysis loses some generality at this point, but the resulting expressions will be more easily interpreted as to physical meaning. The amplitude distribution in the x_i plane a distance z_i from the hologram is given by

$$G(x_i) = \left(\frac{-i}{\lambda z_i}\right)^{1/2} e^{ikz_i} \int_{-\infty}^{\infty} C(x) I_1(x) \exp\left[ik\frac{(x-x_i)^2}{2z_i}\right] dx, \tag{5.112}$$

where we are again assuming that the amplitude transmittance of the hologram is equal to $I_1(x)$ and we are neglecting terms of the order of $1/z_i^2$ in the binomial expansion of the distance between a point x in the hologram plane and the point x_i in the x_i plane. Substitution of (5.110) and (5.111) into (5.112) yields

$$G(x_i) = \frac{-i}{2\pi}\left[\frac{-iR'}{\lambda z_i z_o}\right]^{1/2} e^{ik(z_i - z_o)} \int_{-\infty}^{\infty} \exp\left(-i\frac{k}{2R}x^2\right) \exp\left[i\frac{k}{2z_i}(x - x_i)^2\right]$$

$$\times \int_{-\infty}^{\infty} \tilde{S}\left(-\frac{\gamma}{2\pi} + \frac{\sin\alpha_R}{\lambda}\right) \exp\left[-\frac{R'}{2k}\left(1 + \frac{lR'}{z_o^2}\right)\gamma^2\right] e^{i\gamma x}$$

$$\times \int_{-\infty}^{\infty} F(x_o) \exp\left[i\frac{k}{2l}\left(\frac{lR'}{kz_o}\gamma - x_o\right)^2\right] d\gamma\, dx_o\, dx. \tag{5.113}$$

Now by writing $z_i = z_o$, so that we obtain the primary image, and using the fact that

$$\int_{-\infty}^{\infty} e^{-i(k/2R')x^2} \exp\left[-ix\left(\frac{kx_i}{z_o} - \gamma\right)\right] dx$$

$$= (-i\lambda R')^{1/2} \exp\left[i\frac{R'}{2k}\left(k\frac{x_i}{z_o} - \gamma\right)^2\right], \tag{5.114}$$

we find, after some simplification,

$$G_I(x_i) = \frac{R'}{z_o} e^{-i(k/2l)x_i^2} \exp\left(-i\frac{kR'}{z_o} x_i \sin \alpha_R\right)$$

$$\times \iint_{-\infty}^{\infty} \tilde{S}(\nu) F(x_o) e^{i(k/2l)x_o^2} \exp\left(-i\frac{kR'}{z_o} x_o \sin \alpha_R\right)$$

$$\times \exp\left(i\frac{R'}{z_o} 2\pi\nu(x_i + x_o)\right) d\nu \, dx_o. \tag{5.115}$$

This expression tells us that, in general, when the reference wave is spherical and the object is close to the hologram plane, both the resolution and the field of view are restricted by the MTF of the recording medium. This case represents a situation midway between the situations indicated by Eqs. 5.70 and 5.93. In (5.70), for the Fourier transform hologram, the image resolution is independent of the object frequency, but the extent over which the image exists is limited by the MTF. In (5.93) the extent of the image is independent of the MTF, but the resolved object frequencies extend only up to $\nu_{omax} = \nu_c - \sin \alpha_R/\lambda$, that is, the recording medium cutoff frequency less the carrier frequency.

To see more clearly the effect of limited MTF in the present situation, let us again consider an idealized transfer function for the medium:

$$\tilde{S}(\nu) = 1, \qquad -\nu_c \le \nu \le \nu_c$$

$$= 0 \qquad \text{otherwise.} \tag{5.116}$$

Then

$$G(x_i) = \frac{R'}{z_o} e^{-i(k/2l)x_i^2} \exp\left(-i\frac{kR'}{z_o} x_i \sin \alpha_R\right) \int_{-\infty}^{\infty} \int_{-\nu_c}^{\nu_c} F(x_o) e^{i(k/2l)x_o^2}$$

$$\times \exp\left(-i\frac{kR'}{z_o} x_o \sin \alpha_R\right) \exp\left[i\frac{R'}{z_o} 2\pi\nu(x_i + x_o)\right] d\nu \, dx_o$$

$$= \frac{2\nu_c R'}{z_o} e^{-i(k/2l)x_i^2} \exp\left(-i\frac{kR'}{z_o} x_i \sin \alpha_R\right)$$

$$\times \int_{-\infty}^{\infty} \text{sinc}\left[\frac{2\pi R'}{z_o}(x_i + x_o)\nu_c\right] F(x_o) e^{i(k/2l)x_o^2}$$

$$\times \exp\left(-i\frac{kR'}{z_o} x_o \sin \alpha_R\right) dx_o. \tag{5.117}$$

The line spread function for the imaging process may be found by writing $F(x_o) = \delta(x_o)$. We then obtain

$$|G_I(x_i)|^2 = 4 \left(\frac{R'}{z_o}\right)^2 v_c^2 \text{sinc}^2 \left(\frac{2\pi R'}{z_o} x_i v_c\right)$$

$$= 4 \left(\frac{R}{z_o - R}\right)^2 v_c^2 \text{sinc}^2 \left(\frac{2\pi R}{z_o - R} x_i v_c\right), \qquad (5.118)$$

which has an approximate width

$$\Delta x_i = \left(\frac{z_o - R}{R}\right) \frac{1}{v_c}. \qquad (5.119)$$

An interesting special case is $R = z_o$, that is, reference source in the object plane. Returning to (5.103), we see that the substitution $R = z_o$ eliminates the exponential which is quadratic in $(x - u)$. Thus the primary image term becomes

$$I_1(x) = \left(\frac{i}{\lambda z_o}\right)^{\frac{1}{2}} e^{-ikz_o} \int_{-\infty}^{\infty} F(x_o) e^{-i(k/2z_o)x_o^2}$$

$$\times \int_{-\infty}^{\infty} S(u) \exp\left[ik(x - u)\left(\frac{x_o}{z_o} + \sin \alpha_R\right)\right] du \, dx_o. \quad (5.120)$$

Again, we note that this is a convolution integral between $S(u)$ and $\beta(u)$, where

$$\beta(u) = e^{iku[(x_o/z_o) - \sin \alpha_R]}. \qquad (5.121)$$

Without the quadratic exponential, we find

$$\tilde{\beta}(v) = \delta\left(\frac{x_o}{\lambda z_o} + \frac{\sin \alpha_R}{\lambda} - v\right) \qquad (5.122)$$

so that

$$I_1(x) = \left(\frac{i}{\lambda z_o}\right)^{\frac{1}{2}} e^{-ikz_o} e^{ikx \sin \alpha_R}$$

$$\times \int_{-\infty}^{\infty} F(x_o) \tilde{S}\left[\frac{\sin \alpha_R}{\lambda} + \frac{x_o}{\lambda z_o}\right] e^{ikx_o^2/2z_o} e^{ikxx_o/z_o} \, dx_o. \quad (5.123)$$

Illuminating with a wave $C(x)$ given by (5.111), the primary image is given by

$$G(x_i) = \left(\frac{-i}{\lambda z_i}\right)^{\frac{1}{2}} \int_{-\infty}^{\infty} C(x) I_1(x) e^{ikz_i} e^{i(k/2z_i)(x - x_i)^2} \, dx, \qquad (5.124)$$

where we are again assuming that the amplitude transmittance of the processed hologram is simply equal to the incident irradiance. Substituting (5.111) and (5.123) into (5.124) and letting $z_i = z_o$ as before, we find that

$$G(x_i) = e^{i(k/z_o)x_o^2}F(x_i)\tilde{S}\left(\frac{\sin \alpha_R}{\lambda} + \frac{x_i}{\lambda z_o}\right). \qquad (5.125)$$

The similarity between (5.125) and (5.70) indicates why this type of holo-gram is referred to as a lensless Fourier transform hologram. The MTF of the recording medium, $\tilde{S}(\nu)$, affects only the field of view and not the resolution in the image. The same comments that applied to (5.70) also apply to (5.125).

5.4 THIRD-ORDER ABERRATIONS [4]

5.4.1 Introduction

This section will treat the third-order aberrations of reconstructed wave-fronts. Aberrations are introduced whenever any of the system parameters are changed between recording and reconstructing, such as wavelength, radii of curvature, or even the hologram itself. There will be no discussion of the relationship between the amount of aberration present and image resolution, because this is an enormously complicated subject in itself. The five Seidel aberrations, spherical, coma, astigmatism, field curvature, and distortion, all present in a general holographic system, will be derived as a phase difference between the reconstructed wavefront of a point object and a reference sphere.

5.4.2 Analysis

Consider a point object situated at $P_o(x_o, y_o, z_o)$ in our usual coordinate system (Fig. 5.13). Let the wavelength be λ_o; then the phase of the spherical wave from P_o in the hologram, relative to the phase at the origin, is

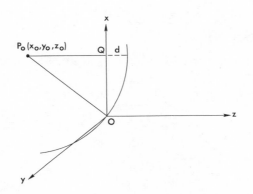

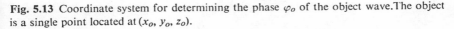

Fig. 5.13 Coordinate system for determining the phase φ_o of the object wave. The object is a single point located at (x_o, y_o, z_o).

$$\varphi_o(x, y) = \frac{2\pi}{\lambda_o} d = \frac{2\pi}{\lambda} \overline{(P_oQ - P_oO)}$$

$$= \frac{2\pi}{\lambda_o} \{[(x - x_o)^2 + (y - y_o)^2 + z_o^2]^{1/2} - (x_o^2 + y_o^2 + z_o^2)^{1/2}\}$$

$$= \frac{2\pi}{\lambda_o} z_o \left\{ \left[1 + \frac{(x - x_o)^2 + (y - y_o)^2}{z_o^2} \right]^{1/2} - \left(1 + \frac{x_o^2 + y_o^2}{z_o^2} \right)^{1/2} \right\}.$$

$$(5.126)$$

Assuming that $z_o^2 > x_o^2 + y_o^2$, we may expand the square roots to obtain

$$\varphi_o(x, y) = \frac{2\pi}{\lambda_o} \left[\frac{1}{2z_o} (x^2 + y^2 - 2xx_o - 2yy_o) - \frac{1}{8z_o^3} (x^4 + y^4 + 2x^2 y^2 \right.$$

$$- 4x^3 x_o - 4y^3 y_o - 4x^2 yy_o - 4xy^2 x_o + 6x^2 x_o^2$$

$$+ 6y^2 y_o^2 + 2x^2 y_o^2 + 2y^2 x_o^2 + 8xyx_o y_o - 4xx_o^3$$

$$\left. - 4yy_o^3 - 4xx_o y_o^2 - 4xx_o^2 y_o) + \text{higher order terms} \right].$$

$$(5.127)$$

If a point source at (x_R, y_R, z_R) supplies the reference wave, also at λ_o, then a similar expression holds for the phase of the reference wave φ_R, with (x_o, y_o, z_o) replaced by (x_R, y_R, z_R). Similarly, if a point source at (x_c, y_c, z_c) supplies the illuminating wave, its phase is given by the same expression with the substitution of (x_c, y_c, z_c) for (x_o, y_o, z_o), and λ_c, the illuminating wavelength, for λ_o. The phases of the primary and conjugate image waves are given by, respectively,

$$\Phi_p = \varphi_c + \varphi_o - \varphi_R \quad \text{and} \quad \Phi_c = \varphi_c + \varphi_R - \varphi_o. \quad (5.128)$$

5.4.2.1 Magnification

It is logical to discuss magnification here since this is calculated from the first-order terms of (5.127).

To first order in $1/z$, then, the phase of the primary image wave is given by

$$\Phi_p^{(1)} = \frac{2\pi}{\lambda_c} \frac{1}{2z_c} (x^2 + y^2 - 2xx_c - 2yy_c)$$

$$+ \frac{2\pi}{\lambda_o} \frac{1}{2z_o} (x'^2 + y'^2 - 2x'x_o - 2y'y_o)$$

$$- \frac{2\pi}{\lambda_o} \frac{1}{2z_R} (x'^2 + y'^2 - 2x'x_R - 2y'y_R). \quad (5.129)$$

The primed coordinates refer to the coordinates of the hologram as it was recorded. A subsequent scaling up or down of the hologram results in the transformation $x = mx'$ and $y = my'$. We now define the illuminating to recording wavelength ratio $\lambda_c/\lambda_o = \mu$ and write

$$\Phi_p^{(1)} = \frac{2\pi}{\lambda_c} \frac{1}{2} \left\{ (x^2 + y^2)\left[\frac{1}{z_c} + \frac{\mu}{m^2 z_o} - \frac{\mu}{m^2 z_R} \right] - 2x\left[\frac{x_c}{z_c} + \frac{\mu x_o}{m z_o} - \frac{\mu x_R}{m z_R} \right] \right.$$

$$\left. - 2y\left[\frac{y_c}{z_c} + \frac{\mu y_o}{m z_o} - \frac{\mu y_R}{m z_R} \right] \right\}. \tag{5.130}$$

We now consider (5.130) to be the first-order term of the Gaussian reference sphere defined by

$$\Phi_p^{(1)} = \frac{2\pi}{\lambda_c} \frac{1}{2} \left[\frac{x^2 + y^2 - 2xa_p - 2yb_p}{Z_p} \right] \tag{5.131}$$

with Z_p its radius and a_p and b_p the coordinates of its center, which determine the Gaussian image point. Making this necessary identification of terms, we obtain

$$Z_p = \frac{m^2 z_c z_o z_R}{m^2 z_o z_R + \mu z_c z_R - \mu z_c z_o} \;;\quad a_p = \frac{m^2 x_o z_o z_R + \mu m x_o z_c z_R - \mu m x_R x_c x_o}{m^2 z_o z_R + \mu z_c z_R - \mu z_c z_o}$$

$$\tag{5.132}$$

with b_p given by an expression similar to that for a_p, but with the x's replaced by y's. The coordinates for the conjugate image are given by

$$Z_c = \frac{m^2 z_c z_o z_R}{m^2 z_o z_R - \mu z_c z_R + \mu z_c z_o} \;;\quad a_c = \frac{m^2 x_c z_o z_R - \mu m x_o z_c z_R + \mu m x_R z_c z_o}{m^2 z_o z_R - \mu z_c z_R + \mu z_c z_o}.$$

$$\tag{5.133}$$

If Z_p is negative, the center of the reference sphere is to the left of the hologram and the image is virtual. The phase Φ_p then corresponds to a wave diverging from Z_p. If it is positive, the primary image is real and to the right of the hologram. The phase $\Phi_p^{(1)}$ then represents a spherical wave converging to Z_p. The same considerations apply to the conjugate image, so that either image may very well be real or virtual.

The magnification is given by

$$M = \frac{da}{dx_o} \tag{5.134}$$

so that

$$M_p = \frac{m}{1 + m^2 z_o/\mu z_c - z_o/z_R}.$$ (5.135)

and

$$M_c = \frac{m}{1 - m^2 z_o/\mu z_c - z_o/z_R}.$$ (5.136)

If $z_c \rightarrow \infty$, that is, plane wave illumination is used, then the μ-dependence of the magnification drops out. Thus in order to achieve magnification by use of different wavelengths, we must use a spherical illuminating wave. Also, if $z_c \rightarrow \infty$, $M_p = M_c$.

The angular magnification for a virtual image is given by

$$M_{\text{ang}} = \frac{d(a/\mathbf{Z})}{d(x_o/z_o)} = \pm \frac{\mu}{m}$$ (5.137)

where the upper sign refers to the virtual image formed by $\Phi_p^{(1)}$ and the lower sign to the virtual image formed by $\Phi_c^{(1)}$. This latter image, then, will always be inverted and the former will always be upright. If the wavelength and scale remain unchanged between recording and illuminating, the angular magnification will be unity, regardless of the other parameters, z_o, z_R, and z_c.

The longitudinal magnification of the holographic imaging process is given by

$$M_{\text{long}} = \frac{d\mathbf{Z}_c}{dz_o} = -\frac{1}{\mu} M^2$$ (5.138)

which, except for the $1/\mu$ factor, is identical to the one for conventional imaging.

5.4.2.2 Third-Order Abberations

The third-order term of the Gaussian reference sphere is given by

$$\Phi^{(3)} = \frac{2\pi}{\lambda_c} \left[-\frac{1}{8} \frac{1}{\mathbf{Z}^3} (x^4 + y^4 + 2x^2 y^2 - 4x^3 a - 4y^3 b - 4xy^2 a \right.$$
$$- 4x^2 yb + 6x^2 a^2 + 6y^2 b^2 + 2x^2 b^2 + 2y^2 a^2 + 8xyab$$
$$\left. - 4xa^3 - 4yb^3 + 4xab^2 - 4ya^2 b) \right].$$ (5.139)

The third-order term of the reconstructed wavefront is given by the sum of the third-order terms of Φ_p and Φ_c as determined from (5.128). Changing to polar coordinates ρ and θ defined by

$$\rho^2 = x^2 + y^2; \quad x = \rho \cos \theta; \quad y = \rho \sin \theta,$$ (5.140)

the third-order aberrations, being the phase differences between the reference sphere and the actual wavefront, separate into the five usual types:

$$\Delta\Phi = \frac{2\pi}{\lambda_c} \ [- \tfrac{1}{8}\rho^4 S \qquad\qquad\qquad\qquad \text{(spherical)}$$

$$+ \tfrac{1}{2}\rho^3(C_x \cos\theta + C_y \sin\theta) \qquad\qquad \text{(coma)}$$

$$- \tfrac{1}{2}\rho^2(A_x \cos^2\theta + A_y \sin^2\theta + 2A_x A_y \cos\theta \sin\theta) \quad \text{(astigmatism)}$$

$$- \tfrac{1}{4}\rho^2 F \qquad\qquad\qquad\qquad \text{(field curvature)}$$

$$+ \tfrac{1}{2}\rho(D_x \cos\theta + D_y \sin\theta)]. \qquad\qquad \text{(distortion)}$$

$$(5.141)$$

Table 5.1 lists the various aberration coefficients in terms of the system parameters.

Table 5.2 gives the same coefficients for the case $z_R = z_c = \infty$, that is, plane reference and illuminating waves.

Table 5.3 again gives the coefficients but for $z_R = z_o$, that is, reference and object points in the same plane, the so-called "lensless" Fourier transform case.

The coefficients listed in these tables are for the conjugate wavefront Φ_c; the aberrations of Φ_p are obtained simply by changing the signs of z_o and z_R. In all three cases we have listed only the x-coefficient for the off-axis aberrations; the y-coefficients are obtained by replacing x with y. Finally, in all expressions for D_x, we have assumed $y_c = y_R = 0$ with no loss of generality.

From Table 5.2 we see that spherical aberration is zero only if $\mu = m$, which means that the hologram is scaled according to the wavelength ratio μ. Spherical aberration disappears completely ($S = 0$) for $z_R = z_o$ as can be seen from Table 5.1.

When $z_c = z_R = \infty$, coma may be minimized by making $\tan \alpha_c = -\mu/m \tan \alpha_R$. Coma will be zero then only for $\mu = m$, just as for spherical. From Table 5.3 we see that for $z_o = z_R$, coma can be eliminated by making $z_c = \pm m z_o$.

From Table 5.2, we see that astigmatism will be minimized for $\tan \alpha_c = -\mu/m \tan \alpha_R$, just as for coma, and A_x will disappear completely if $\mu = m$.

For $z_R = z_o$ (Table 5.3), we see that for $x_c/z_c = -(\mu/m)(x_R/z_R)$ the expression for A_x becomes

$$A_x = \frac{1}{z_o{}^3} \frac{\mu}{m^2} \left[\left(\frac{1}{z_o} + \frac{\mu}{z_c} \right) \left(x_R{}^2 - x_o{}^2 \right) \right]. \qquad (5.142)$$

This can only be made zero for $z_c = \mu z_o$, but it can only be done simultaneously with the condition for zero coma ($z_c = \pm m z_o$) if $\mu = m$.

Table 5.1

Spherical

$$S = \frac{\mu}{m^4}\left[\left(\frac{\mu^2}{m^2}-1\right)\left(\frac{1}{z_o^3}-\frac{1}{z_R^3}\right)-\frac{3\mu}{z_c}\left(\frac{1}{z_o^2}+\frac{1}{z_R^2}\right)+3\left(\frac{m^2}{z_c^2}-\frac{\mu}{m^2 z_o z_R}\right)\left(\frac{1}{z_o}-\frac{1}{z_R}\right)+6\frac{\mu}{z_o z_R z_c}\right]$$

Coma

$$C_x = \frac{\mu}{m}\frac{1}{z_c^2}\left(\frac{x_o}{z_o}-\frac{x_R}{z_R}\right)-\frac{\mu}{m^3}\frac{1}{z_o^2}\left[\frac{x_o}{z_o}\left(1-\frac{\mu^2}{m^2}\right)+\frac{\mu}{m}\frac{x_c}{z_c}+\frac{\mu^2}{m^2}\frac{x_R}{z_R}\right]+\frac{\mu}{m^3}\frac{1}{z_R^2}\left[\frac{x_R}{z_R}\left(1-\frac{\mu^2}{m^2}\right)-\frac{\mu}{m}\frac{x_c}{z_c}+\frac{\mu^2}{m^2}\frac{x_o}{z_o}\right]$$
$$+2\frac{\mu}{m^2}\left(\frac{x_c}{z_c}-\frac{\mu}{m}\frac{x_o}{z_o}+\frac{\mu}{m}\frac{x_R}{z_R}\right)\left(\frac{1}{z_o z_c}-\frac{1}{z_c z_R}+\frac{\mu}{m^2}\frac{1}{z_o z_R}\right)$$

Astigmatism

$$A_x = \frac{\mu}{m^2}\frac{x_c^2}{z_c^2}\left(\frac{1}{z_o}-\frac{1}{z_R}\right)-\frac{\mu}{m^2}\frac{x_o^2}{z_o^2}\left[\frac{1}{z_o}\left(1-\frac{\mu^2}{m^2}\right)+\frac{\mu}{m}\frac{x_c}{z_c}+\frac{\mu^2}{m^2}\frac{1}{z_R}\right]+\frac{\mu}{m^2}\frac{x_R^2}{z_R^2}\left[\frac{1}{z_R}\left(1-\frac{\mu^2}{m^2}\right)-\frac{\mu}{m}\frac{1}{z_c}+\frac{\mu^2}{m^2}\frac{1}{z_o}\right]$$
$$+2\frac{\mu}{m}\left(\frac{1}{z_c}-\frac{\mu}{m^2}\frac{1}{z_o}+\frac{\mu}{m^2}\frac{1}{z_R}\right)\left(\frac{x_o x_c}{z_o z_c}-\frac{x_c x_R}{z_c z_R}+\frac{\mu}{m}\frac{x_o x_R}{z_o z_R}\right)$$

Field curvature

$$F = A_x + A_y$$

Distortion

$$D_x = \frac{\mu}{m}\left[\left(\frac{\mu^2}{m^2}-1\right)\left(\frac{x_o^3}{z_o^3}-\frac{x_R^3}{z_R^3}+\frac{x_o y_o^2}{z_o^3}\right)+\frac{3x_o}{z_o}\left(\frac{x_c}{z_c}+\frac{\mu}{m}\frac{x_R}{z_R}\right)^2-\frac{\mu}{m}\frac{(3x_o^2+y_o^2)}{z_o^2}\left(\frac{x_c}{z_c}+\frac{\mu}{m}\frac{x_R}{z_R}\right)-3\frac{x_c x_R}{z_c z_R}\left(\frac{x_c}{z_c}+\frac{\mu}{m}\frac{x_R}{z_R}\right)\right]$$

Table 5.2

Spherical

$$S = \frac{\mu}{m^4} \frac{1}{z_o^3} \left(\frac{\mu^2}{m^2} - 1 \right)$$

Coma

$$C_x = \frac{\mu}{m^3} \frac{1}{z_o^2} \left[\frac{x_o}{z_o} \left(\frac{\mu^2}{m^2} - 1 \right) - \frac{\mu}{m} \tan\alpha_c - \frac{\mu^2}{m^2} \tan\alpha_R \right]$$

Astigmatism

$$A_x = \frac{\mu}{m^2} \frac{1}{z_o} \left[\frac{x_o^2}{z_o^2} \left(\frac{\mu^2}{m^2} - 1 \right) + \left(\tan\alpha_c + \frac{\mu}{m} \tan\alpha_R \right)^2 - 2 \frac{\mu}{m} \frac{x_o}{z_o} \left(\tan\alpha_c + \frac{\mu}{m} \tan\alpha_R \right) \right]$$

Field curvature

$$F = A_x + A_y$$

Distortion

$$D_x = \frac{\mu}{m} \left\{ \left(\frac{x_o^3}{z_o^3} - \tan^3\alpha_R + \frac{x_o y_o^2}{z_o^3} \right) \left(\frac{\mu^2}{m^2} - 1 \right) + \left(\tan\alpha_c + \frac{\mu}{m} \tan\alpha_R \right) \left[3 \frac{x_o}{z_o} \tan\alpha_c + 3 \frac{\mu}{m} \frac{x_o}{z_o} \tan\alpha_R \right. \right.$$
$$\left. \left. - \frac{\mu}{m} \left(\frac{3x_o^2 + y_o^2}{z_o^2} \right) - \underbrace{3 \tan\alpha_c \tan\alpha_R} \right] \right\}$$

Table 5.3

Spherical

$$S = 0$$

Coma

$$C_x = \frac{\mu}{m}\left(\frac{x_o - x_R}{z_o}\right)\left(\frac{1}{z_c^2} - \frac{1}{m^2 z_o^2}\right)$$

Astigmatism

$$A_x = \frac{\mu}{m}\left(\frac{x_R^2}{z_o^2}\right)\left(\frac{1}{mz_o} - \frac{\mu}{mz_c}\right) + 2\frac{\mu}{m}\frac{1}{z_c z_o}\left(\frac{x_o x_c}{z_c} + \frac{\mu}{m}\frac{x_o x_R}{z_o} - \frac{x_c x_R}{z_c}\right) - \frac{\mu}{m}\frac{x_o^2}{z_o^2}\left(\frac{1}{mz_o} + \frac{\mu}{mz_c}\right)$$

Field curvature

$$F = A_x + A_y$$

Distortion

$$D_x = \frac{\mu}{m}\left[\frac{x_o^3 - x_R^3}{z_o^3}\left(\frac{\mu^2}{m^2} - 1\right) + \frac{x_o y_o^2}{z_o^3}\left(\frac{\mu^2}{m^2} - 1\right) + \left(\frac{x_c}{z_c} + \frac{\mu}{m}\frac{x_R}{z_o}\right)\left(3\frac{x_o x_c}{z_o z_c} + 3\frac{\mu}{m}\frac{x_o x_R}{z_o^2} - 3\frac{\mu}{m}\frac{x_o^2}{z_o^2} - \frac{\mu}{m}\frac{y_o^2}{z_o^2} - 3\frac{x_c x_R}{z_c z_o}\right)\right]$$

All of the remarks applying to astigmatism also pertain to curvature of field, since the defining expressions are identical.

Table 5.1 indicates that distortion is also removed for $x_c/z_c = -(\mu/m)$ (x_R/z_R) and $\mu = m$.

Thus we conclude that all of the primary aberrations will disappear simultaneously by using plane wave reference and illuminating beams of equal but opposite offset angles and by scaling the hologram in the ratio of the wavelengths. It should be remembered that this analysis applies *only* to thin holograms. An aberration-free real image may still be produced with thick holograms, although a convergent illuminating wave is required, as discussed in Chapter 4.

<h1 style="text-align:center">REFERENCES</h1>

[1] R. F. van Ligten, *J. Opt. Soc. Am.*, **56,** 1 (1966).

[2] R. F. van Ligten, *J. Opt. Soc. Am.*, **56,** 1009 (1966).

[3] A. Vander Lugt and R. H. Mitchel, *J. Opt. Soc. Am.*, **57,** 372 (1967).

[4] R. W. Meier, *J. Opt. Soc. Am.*, **56,** 219 (1966).

6 Practical Considerations

6.0 LIGHT SOURCES FOR HOLOGRAPHY

6.0.1 The Gas Laser

By far the most common and important light source for holography today is the gas laser, therefore it will be discussed first. We will discuss the theory of operation of gas lasers only as far as required for the production of good holograms. For an excellent, comprehensive review of gas lasers, the reader is referred to the article by Bloom [1], which contains a large list of references.

Of the many types of gas lasers presently available, the He-Ne laser is the one used for most of the hologram work being performed today, although the ionized argon laser is being used more and more. Therefore most of this discussion will center around the He-Ne laser emitting at .6328 μ, and the argon laser emitting principally at .4880 μ and .5145 μ. The general features of the lasers which will be discussed here are applicable to other gas lasers, since most of the characteristics of interest for holography pertain to properties of the resonator cavity rather than to the physics of the laser action. Of primary interest will be the temporal and spatial properties of the light.

6.0.1.1 Temporal Coherence

A gas laser basically consists of an atomic or molecular gas at low pressure contained in a long discharge tube. The axis of the discharge tube defines the optical axis of the system. Centered on this axis, at each end of the tube, is a high-reflectivity mirror, aligned so that light traveling along the axis reflects back and forth many times through the amplifying medium. A small percentage of this light is transmitted by one or both mirrors on each pass. This transmission represents a loss to the system, therefore for relatively low-gain lasers such as the He-Ne, high-quality, high-reflectivity mirrors are required. For higher gain lasers, such as the argon laser, the mirrors may be allowed to transmit much more light, resulting in a higher output power.

The gas is excited by means of an electrical discharge, either at radio frequencies, dc, or pulsed. Under certain conditions a population inversion (i.e., more atoms in a higher lying level than in a lower level) may be made to exist. Under these conditions the gas discharge becomes a light amplifier for the transition for which there is an inversion. A spontaneous emission causes stimulated emission and the process cascades as the light travels along the tube. Because of the inversion, amplification of the spontaneous emission results. As the light reflects back and forth between the mirrors, the random nature of the spontaneous emissions is swamped by the in-phase, coherent nature of the stimulated emissions. The temporal coherence of the light output is quite high, but because of the nature of the amplifying transition, most laser beams are not single, but multiple frequency. Let us first examine how this multifrequency (multimode) output affects holography.

Because of the low pressure of the gas in a gas laser, the transition of interest is always Doppler broadened. This line, which would normally be an absorption line, becomes the gain envelope when an inversion exists. Thus the laser is a resonant Fabry-Perot cavity with gain. The condition for resonance is that the round-trip path between the mirrors be an integral number of wavelengths:

$$2d = m\lambda, \tag{6.1}$$

where d is the distance between the laser mirrors (cavity length), λ the wavelength at Doppler line center, and m is an integer, the axial mode number. Since $d \gg \lambda$, it is possible that the separation between adjacent allowed resonances (axial modes) may be less than the Doppler line-width (gain envelope). If the gain is sufficiently high, more than one mode may oscillate at once, yielding a multifrequency output. The axial mode separation is determined by calculating what change in wavelength changes the axial mode number by unity:

$$\Delta\lambda_c = \frac{\lambda^2}{2d}, \tag{6.2}$$

where $\Delta\lambda_c$ is the wavelength separation between cavity modes. Thus, if the laser is oscillating in a mode of wavelength λ, the adjacent allowable modes are at $\lambda \pm \Delta\lambda_c$. A typical laser spectrum is shown in Fig. 6.1. Note that for the situation shown, only three modes will oscillate since the allowable

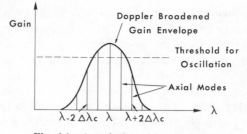

Fig. 6.1 A typical gas laser spectrum.

modes at $\lambda \pm 2\Delta\lambda_c$ fall below threshold. If the gain were to be increased, these are the next modes which would begin to oscillate. Because of the relation (6.2), it is possible to obtain single mode operation by shortening the laser cavity. For the He-Ne laser, $\lambda = .6328\ \mu$ and the Doppler width is of the order of $1.8 \times 10^{-6}\ \mu$ so that the axial mode separation equals the Doppler width for a cavity length of about 10 cm. The He-Ne system possesses sufficient gain so that single mode lasers of this length can be made. The power output will be relatively small, however, and single-mode operation does not imply single-frequency operation, since the single mode may wander about under the gain envelope. This wandering may be caused by thermally induced changes in the cavity length, for example. In order to obtain more useful power levels, longer lasers are generally employed, resulting in a multimode output.

Since holography is basically a two-beam interference problem, consider the simple experiment shown schematically in Fig. 6.2. The laser cavity is formed by mirrors M_1 and M_2. The light transmitted by M_2 is split into two beams at the beamsplitter S, one of which travels to the point P on the recording medium; the other is reflected by the mirror R to the point P. The angle between the two beams is denoted by φ. These two beams will interfere at the plate in the x-y plane, forming straight line fringes lying in the x-direction. Exposure of these fringes requires a finite exposure time,

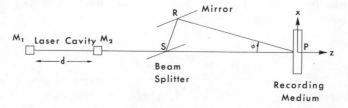

Fig. 6.2 A simple two-beam interference experiment for describing the effects of using a multimode laser for holography.

however, and they must remain steady during this time. In general, a time-dependent wavelength from the laser, a time-dependent optical path difference $D = \overline{SRP}\text{-}\overline{SP}$, or a stable, but multimode laser will cause demodulation of the fringes at P. Figure 6.3 shows two plane wavefronts interfering at P. For small angles, $\sin \varphi = \varphi$, and we can write the phase of wave (1) in the x-direction as

$$\delta_1(x) = kx\frac{\varphi}{2} \qquad (6.3)$$

where $k = 2\pi/\lambda$. Similarly, for wave (2) we have

$$\delta_2(x) = -kx\frac{\varphi}{2}. \qquad (6.4)$$

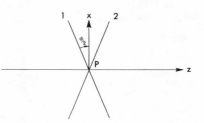

Fig. 6.3   A schematic of the wavefronts of two interfering plane waves.

In the arrangement of Fig. 6.2, one of the waves will have traveled an extra distance D, and the phase difference along x will be

$$\delta(x) = k\,(D + x\varphi). \qquad (6.5)$$

A bright fringe will be produced at x when $\delta(x)$ is an integral multiple of 2π. This leads to the condition

$$D + x\varphi = m\lambda \qquad \text{(bright fringe)} \qquad (6.6)$$

Here m is the order of interference; its average value is D/λ. Now suppose we make a small change in $\Delta\lambda$ in the wavelength. The fringes will shift an amount Δx given by

$$\Delta x = m\frac{\Delta\lambda}{\varphi}. \qquad (6.7)$$

Writing $m = D/\lambda$, this becomes

$$\Delta x = \Delta\lambda\frac{D}{\lambda\varphi}. \qquad (6.8)$$

But $\varphi = \lambda\nu_f$ relates the angle between the beams, the wavelength and the spatial frequency of the fringes, ν_f, so writing $\Delta x_f = 1/\nu_f$ for the fringe spacing, we have

$$\Delta x = \frac{\Delta\lambda}{\lambda^2}D\,\Delta x_f \qquad (6.9)$$

as the distance a fringe moves because of a change in wavelength $\Delta\lambda$. Most commercial lasers used for holography are not completely stabilized, so small thermal variations can cause changes in the cavity length d. A change in d of $\lambda/2$ results in a wavelength shift for each mode of $\Delta\lambda = \Delta\lambda_c$. For a well-constructed laser we can expect thermal drifts to be such that $\lambda \to \lambda + \Delta\lambda_c$ in times of the order of seconds or less. In some lasers the situation is much worse, with the modes jumping about randomly and rapidly. For a change in wavelength $\Delta\lambda = \Delta\lambda_c$, the laser output spectrum is identical to that for $\Delta\lambda = 0$, therefore we need not consider wavelength shifts larger than $\Delta\lambda_c$. The effect of these shifts on the interference pattern can be analyzed as follows: suppose that the path difference between the two beams D is to be expressed as a fraction of the cavity length d:

$$D = Kd. \tag{6.10}$$

Then we can write

$$\Delta x = \frac{\Delta\lambda_c}{\lambda^2} Kd \, \Delta x_f \tag{6.11}$$

which expresses the change in position of a fringe for a wavelength shift $\Delta\lambda_c$ and path difference Kd. But $\Delta\lambda_c = \lambda^2/2d$, so

$$\Delta x = \tfrac{1}{2} K \, \Delta x_f. \tag{6.12}$$

Thus we see that an expansion of the laser cavity by $\lambda/2$ causes the output spectrum to shift by $\Delta\lambda_c$, resulting in a fringe displacement Δx. For a path difference between the two beams equal to the cavity length d ($K = 1$), the fringes will shift by an amount $\Delta x_f/2$. If this happens during one exposure time, the fringe pattern on the photographic plate will essentially be completely demodulated, that is, no fringes will be recorded.

Note that if we express the fringe shift as a fraction of the fringe spacing,

$$\frac{\Delta x}{\Delta x_f} = \tfrac{1}{2} K, \tag{6.13}$$

we see that it is independent of the spatial frequency of the fringe pattern; high-frequency fringes (large φ) are no more sensitive to wavelength variation than are low-frequency fringes.

Next let us assume that the laser is operating at a single wavelength which does not drift in time (single mode, stabilized) and examine the effect on the interference fringes of a time-varying optical path difference. Recall from Fig. 6.2 that the two beams of light travel different paths to the photographic plate. A time-dependent path difference can be introduced by unequal temperature in the two arms which change in time, slight air drafts in the room, or by vibration of the mirror and beam splitter supports. We can analyze these effects by referring to Eq. 6.6:

$$D + x\varphi = m\lambda. \quad \text{(bright fringe)}$$

A small change in D, denoted by ΔD, results in a change in fringe position Δx given by

$$\Delta x = \frac{\Delta D}{\varphi} \qquad (6.14)$$

where we are neglecting the minus sign and assuming that λ and m remain constant. Writing

$$\varphi = \lambda \nu_f = \frac{\lambda}{\Delta x_f} \qquad (6.15)$$

we have

$$\Delta x = \Delta x_f \cdot \frac{\Delta D}{\lambda}. \qquad (6.16)$$

Here we see that we have essentially complete demodulation of the fringes for $\Delta D = \lambda/2$. We also see that the fractional fringe shift, $\Delta x/\Delta x_f$ is independent of the frequency of the fringes. Here again the sensitivity of the modulation of the fringe system to path variations is no greater for high-frequency fringes than for low-frequency fringes.

Next let us examine how the modulation (visibility) of the fringe pattern is affected by a multimode laser. We will assume that both the mode position and the path difference are stationary in time.

A gas laser of sufficient gain oscillates in more than one longitudinal (axial) mode if the separation of these modes is smaller than the width of the atomic line of the transition involved. The wavelength separation of adjacent axial modes is given by (6.2):

$$\Delta \lambda_c = \frac{\lambda^2}{2d}.$$

The modulation M of an interference pattern may be defined as

$$M = \frac{I_{max} - I_{min}}{I_{max} + I_{min}} \qquad (6.17)$$

where I_{max} and I_{min} are, respectively, the values of the irradiance at the maxima and minima of the fringe pattern. This modulation is related to the spectral intensity distribution of the light source through a Fourier transform. A good treatment of this relationship may be found in Born and Wolf [2]. Here it is shown that the modulation of the fringes of two interfering beams of equal irradiance can be written

$$M = \frac{(S^2 + C^2)^{1/2}}{P}, \qquad (6.18)$$

where

$$S(D) \equiv 2 \int j(x) \sin{(xD)} \, dx \qquad (6.19)$$

$$C(D) \equiv 2 \int j(x) \cos{(xD)} \, dx \qquad (6.20)$$

$$P \equiv 2 \int j(x) \, dx. \qquad (6.21)$$

Here D is the optical path difference as before and $j(x)$ is the spectral intensity distribution of the light source. The quantity

$$x = k - k_0 \qquad (6.22)$$

where $k = 2\pi/\lambda$ and $k_0 = 2\pi/\lambda_0$; λ_0 is an arbitrarily chosen wavelength near the center of the Doppler line.

A good approximation to a multimode gas laser source is that $j(x)$ is a sum of Dirac delta functions $\delta(x)$, with each spike separated from its nearest neighbor by

$$\Delta k = \frac{2\pi}{2d}. \qquad (6.23)$$

If we also assume that the spectral distribution is symmetrical about line center, then $S(D) = 0$ and the fringe modulation can be written

$$M = \frac{|C(D)|}{P}. \qquad (6.24)$$

For the delta function spectrum, we have, for a single mode,

$$j_1(x) = \delta(x), \qquad (6.25)$$

for two modes

$$j_2(x) = \delta\left(x - \frac{\Delta k}{2}\right) + \delta\left(x + \frac{\Delta k}{2}\right), \qquad (6.26)$$

and for three modes

$$j_3(x) = \delta(x - \Delta k) + \delta(x) + \delta(x + \Delta k) \qquad (6.27)$$

and so on. Using these expressions for the $j_n(x)$, we obtain for the fringe modulation M_n where n denotes the number of modes oscillating,

$$M_1(D) = 1$$

$$M_2(D) = \frac{\left| 2\cos\left(\frac{\Delta k}{2} D\right) \right|}{2} = \left| \cos\left(\frac{\pi D}{2d}\right) \right|$$

$$M_3(D) = \frac{|2\cos(\Delta kD) + 1|}{3} = \frac{\left| 2\cos\frac{\pi D}{d} + 1 \right|}{3}$$

$$M_4(D) = \frac{\left| 2\cos\left(\frac{3\Delta k}{2} D\right) + 2\cos\left(\frac{\Delta k}{2} D\right) \right|}{4} = \frac{\left| \cos\left(\frac{3\pi D}{2d}\right) + \cos\left(\frac{\pi D}{2d}\right) \right|}{2}$$

$$M_5(D) = \frac{|2\cos(2\Delta kD) + 2\cos(\Delta kD) + 1|}{5}$$

$$= \frac{\left| 2\cos\left(\frac{2\pi D}{d}\right) + 2\cos\left(\frac{\pi D}{d}\right) + 1 \right|}{5}$$

and so on. Evaluating these for path differences equal to an integral multiple of the cavity length yields

$$M_n(0) = 1 \qquad \text{regardless of } n$$

$$M_n(d) = \frac{1}{n} \qquad n \text{ odd}$$

$$= 0 \qquad n \text{ even}$$

$$M_n(2d) = 1 \qquad \text{regardless of } n.$$

Thus for zero path difference the fringe modulation is 100% no matter how many modes are oscillating. For a path difference equal to one cavity length d, the fringe modulation depends on the number of modes. For example, if three modes are oscillating, the modulation is 1/3, assuming all have equal strength. In any actual laser, of course, the modes may not be of equal strength and will have to be weighted accordingly. As a typical example, suppose the three modes are in the ratio 1:2:1. Then $P = 4$ and D ($D = d$) $= 2\cos\Delta kd + 2$, therefore $M_3(d) = 0$ and not 1/3. Thus the modulation of the fringes at $D = d$ will generally be less than $1/n$ for an odd number of modes.

In summary we note that we have examined the problem of forming interference fringes with gas laser light for three cases: (a) a time-varying wavelength in the laser output, (b) a time-varying optical path difference between the two interfering beams, and (c) a multimode, but stable laser output. All three of these perturbations can cause a serious demodulation of the fringes with a resultant loss in diffracted flux, or possibly even a complete loss of the hologram. These effects can occur singly or in any combination. In all cases the degree of demodulation was found to be independent of the spatial frequency of the fringes. Problem (a) can be eliminated by working near zero path difference, (b) by stabilization of the supports and by blocking possible air drafts, and (c) by working near zero path difference or a path difference that is an even multiple of the cavity length.

A nonzero spectral bandwidth of the source used for illuminating the hologram mainly affects the resultant image resolution, as discussed earlier. Because of this the gas laser still represents the optimum illuminating source because of its high power in a narrow bandwidth. Even though the effective bandwidth of a multimode laser must be taken as the full Doppler width, this is still usually narrower than most spectral sources.

6.0.1.2   Spatial Coherence

The spatial coherence of the light used in holography is important in two respects. In recording the hologram, both reference and object beams must have well-defined wavefronts which are constant in time. The object wave may, of course, be highly complex in shape, but should nevertheless be constant. Also, in reconstructing the object wave, the resolution in the final image will depend on the size of the source of the illuminating wave, as discussed in Chapter 5.

The spatial coherence of a light field is a measure of the degree of phase correlation at two different points of space at a single time. This correlation is primarily dependent on the size of the source from which the light originated, and to find this it is necessary to know the precise spatial distribution of flux from a gas laser.

The lowest order (spatial) mode from a gas laser has an irradiance variation of the form

$$I(\rho) = e^{-\rho^2/w^2} \tag{6.28}$$

where ρ is the radial distance from the axis and w is called the spot size; it is the value of ρ at which the amplitude has fallen to $\frac{1}{e}$ of its central value. The wavefront is either spherical or plane. Suppose we wish to focus the laser light with this distribution to a small point with a lens. The spot size at focus is given by

$$w_0 = \frac{\lambda}{\pi \text{ N.A.}} \tag{6.29}$$

where λ is the wavelength of the light and N.A. is the numerical aperture of the lens used to focus the light, which in this case is w/f, f being the focal length of the lens. If w is larger than the diameter of the focusing lens, then the focused spot size will be just the Airy disk radius. If w is smaller than the lens diameter, the irradiance distribution at focus is given by

$$I(\rho) = e^{-\rho^2/w_o^2} = e^{-\pi^2\rho^2(\text{N.A.})^2/\lambda^2} \tag{6.30}$$

For $\lambda = .6328 \, \mu$ (He-Ne laser), the spot size as a function of the N.A. of the focusing objective is

$$w_0 = \frac{.2}{\text{N.A.}} \text{ microns} = \frac{.2f}{w} \text{ microns} \tag{6.31}$$

assuming that the spot size w of the light entering the lens is smaller than the lens diameter. This is the spot size which will determine the resolution in the reconstruction, as described in Chapter 5.

6.0.2 The Solid State Laser

The solid state laser most often used for holographic applications is the pulsed ruby laser, emitting at a wavelength just slightly longer than that of the He-Ne gas laser—$.6943 \, \mu$. This wavelength is approximately at the limit of the long wavelength cutoff for red-sensitized photographic emulsions but is nevertheless still quite useable. Because a ruby laser is pulsed, it has some very definite advantages over a continuous source for some aspects of holography. In particular, because of the short pulse duration, some subject movement can be tolerated during an exposure. A hologram made with a short pulse of light from a ruby laser can record the object at one instant of time and later the resulting three-dimensional image can be studied in detail.

The major drawback to making a hologram with a pulsed laser is the limited coherence length of most solid state lasers. Single axial mode He-Ne lasers are available commercially but have low power output. Even multimode gas lasers can be used with path differences of the order of 10–20 cm, whereas with most pulsed lasers one is limited to path differences of the order of 1 cm. Pulsed ruby lasers can be operated in a single axial mode, thus competing with gas lasers for coherence length, by putting additional reflectors in the laser cavity. Hercher [3] has described this method for obtaining single-mode operation of a Q-switched ruby laser. McClung and Wiener have improved the temporal coherence with axial mode control [4].

Another drawback to the use of the pulsed ruby laser as a source for holography is that the wavefront of the output is not simply defined. The

He-Ne gas laser generally operates in the single, lowest order spatial mode, in so-called TEM_{00} mode, having the irradiance distribution given in (6.28). On the other hand, it is difficult to describe the exact spatial distribution of the output of a pulsed ruby laser. Jacobson and McClung have improved the spatial distribution of a ruby laser but at the expense of total energy in the pulse [5]. In view of the photographic sensitivity problem mentioned previously, this energy loss can become quite serious. They were able to make holograms with such a source, however.

Brooks et al. [6] have successfully produced holograms with a pulsed ruby laser without mode control. This has been done by a careful choice of the optical arrangement to minimize the effects of the relatively low temporal and spatial coherence of ruby laser light. This is accomplished by arranging the holographic system for zero path difference between object and reference beams and also by matching as closely as possible the relative spatial positions of the two wavefronts. When these two conditions are met, the two waves will be coherent with each other and a good hologram will be recorded.

A simple arrangement that has been used successfully with the ruby laser is shown in Fig. 6.4. The zero optical path condition is easily met by equalizing the optical paths in each arm. The spatial distribution is matched in the two arms by expanding both beams to the same size and reflecting each the same number of times. The optical path difference between the two beams increases with increasing distance from the center of the hologram, since the order of interference is increasing. This, as with Gabor holograms, limits the effective aperture of the hologram. Brooks et al. [6] have compensated for this effect by using a stepped retardation plate, such as shown in Fig. 6.5. With this arrangement, the ray passing through the center of each step travels the same optical path as the corresponding ray in the other beam. The disturbing effect of diffraction at the edges of the step are minimized with the use of a diffuse object beam. The use of a diffuser causes a serious spatial mismatch of the two wavefronts, however, resulting in a loss in fringe contrast. Any point on the reference wave must

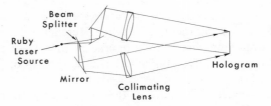

Fig. 6.4 A simple arrangement for spatially matching two interfering waves [6].

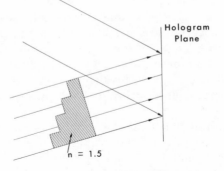

Fig. 6.5   A simple path length compensation scheme [6].

interfere with a point on the object wave whose amplitude and phase will be determined by a sum of components from each point of the original wave. Hence, unless there is a large degree of spatial coherence, the introduction of a diffusing plate into one of the beams will reduce the fringe contrast. For a typical ruby laser, then, only a limited diffusing angle can be recorded.

Brooks et al. have also described a more sophisticated optical arrangement which allows the use of a diffuser with large diffusing angles without seriously upsetting the spatial match. Their arrangement is shown in Fig. 6.6. With this arrangement the diffuser is imaged into the hologram plane. This creates a one-to-one correspondence between a point on the object wavefront and the corresponding point on the reference wavefront. Another lens, identical to the one imaging the diffuser onto the hologram plane (a pair of plano-convex lenses), is placed next to the diffuser and directs

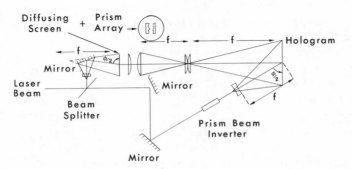

Fig. 6.6   A scheme for matching wavefronts and allowing the use of a diffuser [6].

the diffuse light toward the imaging lens. This lens also compensates for the increasing path lengths at increasing distances from the center of the diffuser. If the reference beam is incident on the hologram plane at an angle $\varphi/2$, then the path length increases from one side of the plane to the other; this is compensated for in the object beam by directing the light onto the diffuser at an angle $\varphi/2$ also. This light is then redirected toward the recording plane by means of an array of small prisms, making optimum use of the light.

This arrangement, although somewhat cumbersome, allows spatial and temporal matching of the two beams to within the limits of the lens aberrations. This is perfectly suitable for making holograms of diffusely illuminated objects with a ruby laser source. It has been tested by Brooks et al. with the object placed midway between the diffuser and the imaging lens. The recorded object was observable over about 40° with reasonably uniform diffuse illumination.

With the foregoing limitations in mind, we will discuss the prime advantage in using a ruby laser as a source for holography—the short pulse duration. Assuming that enough energy to expose the recording medium is emitted in a single pulse, the total holographic exposure is made in a very short time. This means that normal vibration and motion tolerances can be very much relaxed. As a rule of thumb, object motion of about $\lambda/2$ during an exposure will obliterate the hologram because of the resultant motion of the fringes. Since the pulse duration of a typical ruby laser is of the order of 1 msec, the maximum allowable subject velocity is about .35 mm/sec. This is quite an appreciable subject motion when compared with the maximum allowable velocity for a 1-min exposure with a gas laser— 5×10^{-6} mm/sec. But this is still not the ultimate, since sufficient energy for a holographic exposure (on photographic film) can be obtained with a Q-switched ruby laser. Pulse durations of 30 nanosec are typical, and the maximum allowable subject velocity is now as large as 10 m/sec. This number can be further increased by suitable choice of object and geometrical arrangement. An object moving in silhouette, for example, will be subject to motion restrictions essentially the same as in photography; the only portions of the object affected by the motion will be the leading and trailing edges which will suffer a loss in resolution exactly as in conventional photography.

The combination of a Q-switched ruby laser source and holographic interferometry is a very powerful analytic tool. Holographic interferometry will be discussed in greater detail in Chapter 8 and can be described simply as follows. The hologram is double-exposed with two objects. Subsequent illumination reconstructs the two object waves simultaneously, and if the two objects were similar enough (in practice, the same object slightly deformed or displaced is used for the second object) these two waves will interfere and fringes will be visible. With a Q-switched ruby laser source and a diffuse, moving object, a three-dimensional interferogram of the object in motion can be made. Many fine examples of holographic inter-

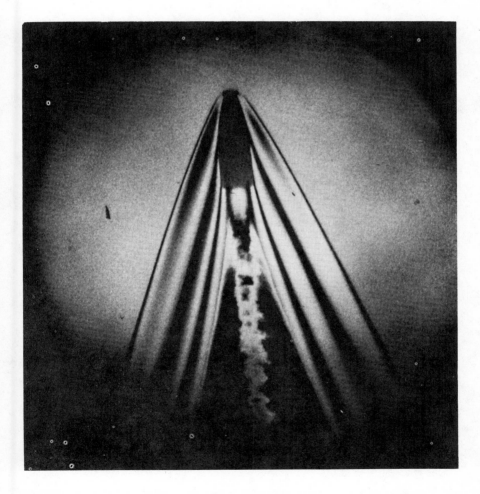

Fig. 6.7   A holographic interferogram of the shock wave of a .22 caliber bullet made with a pulsed ruby laser [7]. (Courtesy of TRW, Inc.)

ferometry with transient phenomena have been demonstrated by Heflinger, Weurker, and Brooks [6, 7]. A few examples of their work are shown in Figs. 6.7, 6.8, and 6.9.

The expense (ruby laser systems are generally more expensive than gas lasers) and limited coherence of the ruby laser have prevented its widespread use as a holographic source. The coherence problems can be overcome with suitable arrangements, and in view of the important advantage of the ultra-short exposure time, the ruby laser should become a very important source for holography.

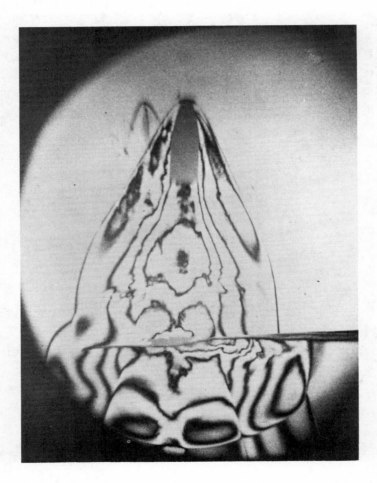

Fig. 6.8   A holographic interferogram of the shock wave of a bullet which has just passed through a sheet of .04mm brass shim stock [7]. (Courtesy of TRW, Inc.)

Fig. 6.9   A holographic interferogram of the change in gas distribution inside a light bulb [7]. (Courtesy of TRW, Inc.)

6.0.3   Line Source

Since the advent of lasers, very few holograms have been made with a nonlaser source. Of course, all of the holograms made before about 1962 were made with nonlaser sources, but the problems were many and the rewards few. Since the cost of a laser is now so small, it is hardly worth the effort to produce holograms with a nonlaser source. But, as mentioned in Chapter 1, the widespread interest in holography today stems not only from the invention of the laser but also from such innovations as the offset reference beam and diffuse illumination. The first off-axis holograms were made with a filtered mercury arc as a source, as were the very first holograms in 1948. The off-axis reference beam hologram requires that some 10^5 interference orders be recorded for a 30° angle between the beams and a 10-cm hologram. Recording this many orders requires a very narrowband source, the maximum allowable bandwidth being only of the order of 0.06 Å. Most arc sources are not this monochromatic and the few which are have a very low output power. There are, however, several methods for reducing the coherence requirements. These are generally interferometer arrangements yielding achromatic fringes [6, 8].

Achromatic fringe systems are formed when the phase difference between the two interfering beams is independent of wavelength. These are formed using a dispersing element such as a grating or prism, as described in Ditchburn [9]. Using such a fringe system, Leith and Upatnieks [10] have produced holograms with a high-pressure mercury arc, although they are of a lesser quality than those made with a laser source.

Their method is best explained using the diffraction grating model for the off-axis type of hologram. Since the hologram represents a modulated diffraction grating, it can be made by imaging a diffraction grating onto the hologram recording medium and modulating this image so that it becomes a hologram. The coherence requirements on the source will be reduced, since an image of the grating will form even in white light.

A suitable arrangement used by them is shown in Fig. 6.10. The second lens images the grating onto plane P_4. Parallel rays diffracted from the grating are imaged at plane P_2. A spatial filter in this plane allows only two orders to pass on to form the grating image. One of these orders is used as a reference beam and the other as an object beam. That the resulting fringe system in plane P_4 is achromatic may be best understood from the viewpoint that the fringes are images of the grating rulings. The fringes can also be considered the result of interference between the two orders passed by the stop in plane P_2. If these are the two first orders, then the recorded fringes will have just twice the spatial frequency of the grating with a sinusoidal irradiance distribution.

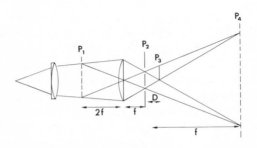

Fig. 6.10   An arrangement for recording an off-axis hologram with a line source [10].

It is easily shown that the required source coherence is least for the in-line holograms first described by Gabor, because the angle between the scattered (object) wave and the background (reference) wave is smaller (for a given size hologram) than for an off-axis arrangement. This small angle implies a small path difference between the two beams at the edges of the hologram and hence less source coherence is required than for an off-axis hologram. For the achromatic fringe system previously described, consider first that the object is placed in one beam in plane P_3 (Fig. 6.10). The waves scattered by the (thin) object interfere with the coherent background and the recorded distribution in plane P_4 reflects in an in-line hologram, with the same requirements on source coherence.

If we now allow another order to be transmitted through plane P_2 to P_4, the wave scattered by the object interferes with this beam to form the grating image. This interference will take place over the same distance along P_4 that the background wave interfered with the scattered wave. But many more fringes (orders) are present because of the larger angle between the two beams. Thus the achromatic fringe hologram method allows off-axis holography with exactly the same source coherence requirements as for in-line holography. All of the usual benefits of off-axis holograms are preserved, such as separation of the conjugate and primary images, relative insensitivity to the nonlinearities of the recording medium, avoidance of the overlapping flare light caused by the self-interference of different object points, and the possibility of using beam balances near unity.

6.1 RECORDING MEDIA FOR HOLOGRAPHY

6.1.1 Resolution and Bandwidth Requirements

6.1.1.1 Recording the Signal [11]

The holographic recording medium must be able to resolve sufficient information so that the object wave may be reconstructed. Just how much resolution is required depends on the object size, type of hologram, and arrangement. In general, an in-line Gabor hologram requires the least amount of bandwidth and an off-axis diffuse-subject hologram the most. For the in-line type, limited resolution of the recording medium results directly in a loss of image resolution. Generally speaking, an off-axis hologram image suffers only a loss of field for limited recording medium resolution—this becomes an exact statement for Fourier transform holograms. The effect of recording medium resolution on image resolution has been discussed in Chapter 5. Here we will limit the discussion only to spatial

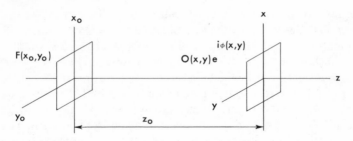

Fig. 6.11 Recording the hologram of a transparency—notation.

frequency bandwidth requirements sufficient to reconstruct the complete object wave, with no loss in image resolution or field.

First we must determine the spatial frequency spectrum of the object. Consider the simple case shown in Fig. 6.11 where the object is a transparency, located a distance z_o from the hologram plane. Assume that the object transparency, with transmittance $F(x_o, y_o)$ is illuminated with a plane, monochromatic light wave of wavelength λ. The field at the hologram plane is given by Eq. 3.2:

$$O(x, y) = \frac{-i}{2\lambda} \iint_A F(x_o, y_o) \, [1 + \cos \theta] \frac{e^{ikr}}{r} \, dx_o \, dy_o,$$

where $k = 2\pi/\lambda$ and θ is always small so we can write $1 + \cos \theta \approx 2$. Replacing r in the denominator by z_o, we have

$$O(x, y) = \frac{-i}{\lambda z_o} \iint_A F(x_o, y_o) \exp \{ik[z^2 + (x - x_o)^2 + (y - y_o)^2]^{1/2}\} \, dx_o \, dy_o.$$

(6.32)

Using the binomial expansion for the square root in the exponent, we have

$$O(x, y) = \frac{-i}{\lambda z_o} \iint_A F(x_o, y_o) \exp\left\{ik\left[z_o + \frac{(x - x_o)^2}{2z_o} + \frac{(y - y_o)^2}{2z_o}\right]\right\} \, dx_o \, dy_o.$$

(6.33)

We can write (6.33) as a convolution

$$O(x, y) = F*g(x, y) \tag{6.34}$$

where

$$g(x, y) = \frac{-i}{\lambda z_o} \exp ik\left[z_o + \frac{x^2}{2z_o} + \frac{y^2}{2z_o}\right]. \tag{6.35}$$

Since a convolution of two functions corresponds to a product of their transforms, we have

$$\tilde{O}(\alpha, \beta) = \tilde{F}(\alpha, \beta) \cdot \tilde{g}(\alpha, \beta) \qquad (6.36)$$

where α and β are spatial frequency variables corresponding to x and y, and

$$\tilde{g}(\alpha, \beta) = \frac{-i}{\lambda z_o} e^{ikz_o} e^{i(z_o/2k)(\alpha^2 + \beta^2)} . \qquad (6.37)$$

Since $|\tilde{g}(\alpha, \beta)| = $ constant, we see that the power spectrum of O and F are the same—both have the same spatial frequency content. What this means is that recording the signal wave requires no more resnlution than does recording the image of the transparency. Since the arrangement of Fig. 6.11 is just the arrangement used to record an in-line hologram, we see that an in-line hologram requires the same resolution of the recording medium as an image-forming system.

The situation is different for an off-axis hologram, since now the diffraction pattern to be recorded is modulated onto a spatial carrier frequency contained in the signal; the resolution requirements on the recording medium will be approximately doubled. Since the interference is generally between two beams defining a plane, the doubling of the resolution requirements is in only one dimension; the requirements for the other dimension remain unchanged.

6.1.1.2 Recording the Field of View

A. FOURIER TRANSFORM HOLOGRAMS. This type of hologram is produced with a plane reference wave and either (a) the object at a very great distance from the hologram plane, or (b) with a lens forming an image of the object at ∞ and the recording medium placed one focal length from the lens. An arrangement for recording a Fourier transform hologram is shown in Fig. 6.12. With this arrangement each point of the object results in a plane

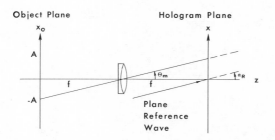

Fig. 6.12   Recording a Fourier transform hologram.

wave at the hologram plane, which interferes with the plane reference wave. If the reference wave makes an angle α_R with the axis, the recorded spatial frequency for a point at the center of the object is

$$\nu_o = \frac{\sin \alpha_R}{\lambda}. \tag{6.38}$$

If the object is of width $2A$, the object point at $x_0 = -A$ results in a plane wave incident on the hologram plane at an angle

$$\theta_{max} = \tan^{-1}\left(\frac{A}{f}\right) \tag{6.39}$$

and the object point at $x_o = A$ yields a plane wave incident at an angle

$$\theta_{max} = -\tan^{-1}\left(\frac{A}{f}\right). \tag{6.40}$$

The range of recorded spatial frequencies is given by

$$\Delta\nu = \nu_{max} - \nu_{min} = \frac{\sin \alpha_R - \sin \theta_{min}}{\lambda} - \frac{\sin \alpha_R - \sin \theta_{max}}{\lambda}$$

$$= \frac{\sin \theta_{max} - \sin \theta_{min}}{\lambda}$$

$$= \frac{2}{\lambda} \sin\left(\tan^{-1}\frac{A}{f}\right). \tag{6.41}$$

Hence we see that for a Fourier transform hologram, the range of spatial frequencies which the recording medium must resolve depends only on the object size—provided, of course, that the hologram is large enough so that the entire field is covered.

B. FRESNEL HOLOGRAMS. Consider the arrangement shown in Fig. 6.13 for recording a Fresnel hologram. Here the object point at $x_o = A$ yields

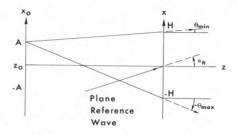

Fig. 6.13 Recording a Fresnel hologram.

rays that are incident on the recording medium of dimension $2H$ over a range of angles

$$\theta_{\min} = \tan^{-1}\left(\frac{H - A}{z_o}\right)$$

$$\theta_{\max} = \tan^{-1}\left(\frac{-H - A}{z_o}\right)$$

(6.42)

with similar expressions for the object point at $-A$. The range of spatial frequencies is given by, for a plane reference wave incident at an angle α_R,

$$\nu_{\max} = \frac{\sin \alpha_R - \sin \theta_{\max}}{\lambda}$$

$$\nu_{\min} = \frac{\sin \alpha_R - \sin \theta_{\min}}{\lambda}$$

and so

$$\Delta\nu = \nu_{\max} - \nu_{\min} = \frac{\sin\left[\tan^{-1}\left(\frac{H - A}{z_o}\right)\right] + \sin\left[\tan^{-1}\left(\frac{H + A}{z_o}\right)\right]}{\lambda}.$$

(6.43)

From this expression we see that the range of spatial frequencies which must be recorded for a Fresnel hologram depends on both object and hologram size.

C. LENSLESS FOURIER TRANSFORM HOLOGRAMS. Although this arrangement was described in Section 3.2.2, we will also include an analysis here. A lensless Fourier transform hologram can be recorded as shown in Fig. 6.14, where we are assuming that an object point and a point source for

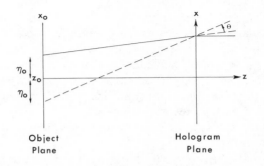

Fig. 6.14 Recording a lensless Fourier transform hologram.

the reference beam are located $+\eta_o$ and $-\eta_o$ from the axis, respectively. The angle θ between the two interfering beams decreases as x increases and therefore the spatial frequency of the exposure variations is decreasing. The angle θ as a function of x is given by an expression analogous to (3.46):

$$\theta(x) = \tan^{-1}\left[\frac{\eta_o + x}{z_o}\right] + \tan^{-1}\left[\frac{\eta_o - x}{z_o}\right]$$

$$= \frac{2\eta_o}{z_o}\left[1 - \frac{1}{3z_o^2}(\eta_o^2 + 3x^2) + \cdots\right]. \qquad (6.44)$$

Since

$$\theta(o) = 2\frac{\eta_o}{z_o}\left[1 - \frac{\eta_o^2}{3z_o^2} + \cdots\right] \qquad (6.45)$$

we see that the range of angles, and hence the spatial frequencies, does not vary rapidly with x. Each object point yields an essentially constant spatial frequency on the recording medium, and we have an approximation to the Fourier transform case.

6.1.2   The Transfer Characteristic of the Recording Medium [12]

6.1.2.1 Introduction

The medium on which the hologram is recorded spatially stores information about the incident irradiance. The transfer characteristic of the medium, that is, the input-output relationship, plays a major role in the holographic process. The primary purpose of the recording medium is, of course, to store the information about the incident irradiance in such a way that the object wavefront can later be reconstructed. This wavefront is then used to form an image and it is this image which we want to optimize. The question immediately arises, "What constitutes the optimum holographic image?" We can imagine a number of criteria such as image sharpness, contrast, brightness, noise, and so forth. The effect of the recording medium on resolution has been discussed in Chapter 5. Rather than trying to solve the complicated problem of image quality here, we will concern ourselves with only one factor—the radiance of the holographic image. This will obviously not represent the end-all for determining the optimum transfer characteristics of the recording medium, but it is an obvious beginning.

The major spatial irradiance detector has for many years been the photographic emulsion, therefore it has become the most commonly used recording medium for holography.   Although other recording media have been suggested and used for holography, the photographic emulsion is still the most generally useful, so we will restrict our discussion here to this detector. The results in general will be applicable to other recording media.   Traditionally photographic films and plates have been used for recording the irradiance of an image, usually as formed by a lens, and as a result they have been optimized with this purpose in mind. Photographic materials, however, have performed surprisingly well as holographic recording media, even though this use represents a considerable departure from that originally intended.

6.1.2.2 Analysis

To begin the analysis, we note that a hologram is a complicated diffracting device; in order to determine how it diffracts light we must know the amplitude transmission characteristics of the emulsion on which it is recorded. We assume that a functional relationship exists between exposure and amplitude transmittance

$$t = f[E(x)].$$

Strictly speaking, the amplitude transmittance t should be expressed as a complex quantity, that is, in terms of a modulus and phase angle.   However, for the purposes of this analysis, we will assume that it is a real quantity equal to the square root of the normally measured irradiance transmittance.   Thus this analysis ignores such factors as photographic relief images and index variations.   Similarly, we will ignore the effects of light scattering in the emulsion, that is, the effects of the modulation transfer function.   Furthermore, the emulsions are assumed to be thin, so that no depth effects occur.

Consider the holographic system of Fig. 6.15a. This is a simple Fourier transform system so that a Fourier transform relationship holds between the plane of the object and the hologram plane H. The light from the object on the plate is represented by $O_o(x)e^{i\varphi_o(x)}$, where $O_o(x)$ is a real function. The reference beam is a plane wave incident on the hologram plane at an angle α_R and is represented by $R_o e^{ikx \sin \alpha_R}$, where R_o is constant and $k = 2\pi/\lambda$. The exposure for unit time at the plate is

$$E(x) = |O_o(x)e^{i\varphi_o(x)} + R_o e^{ikx \sin \alpha_R}|^2$$

$$= R_o^2 + O_o^2(x) + 2R_o O_o(x) \cos [kx \sin \alpha_R - O_o(x)]. \quad (6.46)$$

The term $O_o^2(x)$ is not constant and represents the granular speckle pattern on the photographic plate. We represent it as a constant term O_c plus an oscillating term $O'(x)$. Thus

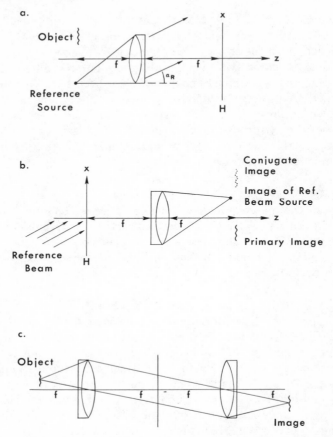

Fig. 6.15 Illustrating the assumed arrangements for describing the effects of the recording medium characteristics. (*a*) Recording a Fourier transform hologram. (*b*) Reconstructing the two waves and forming the two images. (*c*) Forming the direct image.

$$O_o^2(x) = O_c + O'(x).$$

Equation 6.46 can now be separated into similar parts so that

$$E(x) = E_o + F(x) \qquad (6.47)$$

where

$$E_o = R_o^2 + O_c$$

and

$$F(x) = O'(x) + 2R_oO(x) \cos [kx \sin \alpha_R - \varphi_o(x)].$$

When the recorded hologram is illuminated with the reference wave as in Fig. 6.15*b*, there will be two images formed in the transform plane of the second lens. At this point we note only that the developed plate has an amplitude transmittance $t(x)$ which varies spatially about an average transmittance $\langle t(x) \rangle$. If $T(\omega)$ is the Fourier transform of $t(x)$, then by Parseval's theorem,

$$\int_{-H}^{H} |t(x)|^2 \, dx = \frac{1}{2\pi} \int_{-\infty}^{\infty} |T(\omega)|^2 \, d\omega, \qquad (6.48)$$

where we have assumed that the hologram plate extends from $-H$ to H. The right-hand integral over the image plane is the total light flux through this plane.

The average transmittance is given by

$$\langle t(x) \rangle = \frac{1}{2H} \int_{-H}^{H} t(x) \, dx = \text{constant} \qquad (6.49)$$

and by Parseval's theorem,

$$\int_{-H}^{H} |\langle t(x) \rangle|^2 \, dx = 2H|\langle t(x) \rangle|^2 = \frac{1}{2\pi} \int_{-\infty}^{\infty} |T_o(\omega)|^2 \, d\omega, \qquad (6.50)$$

where $T_o(\omega)$ is the Fourier transform of $\langle t(x) \rangle$. Physically, the last integral represents the total specularly transmitted flux.

The variance of $t(x)$ is given by

$$V[t(x)] \equiv \frac{1}{2H} \int_{-H}^{H} |t(x) - \langle t(x) \rangle|^2 \, dx$$

$$= \frac{1}{2H} \int_{-H}^{H} [|t|^2 - t\langle t^* \rangle - \langle t \rangle t^* + |\langle t \rangle|^2] \, dx$$

$$= \frac{1}{2H} \int_{-H}^{H} [|t|^2 - |\langle t \rangle|^2] \, dx$$

where we have used (6.49) to arrive at the last expression. Using (6.48) and (6.50), we can write this as

$$V[t(x)] = \frac{1}{4\pi H} \int_{-\infty}^{\infty} |T(\omega)|^2 \, d\omega - \frac{1}{4\pi H} \int_{-\infty}^{\infty} |T_o(\omega)|^2 \, d\omega. \qquad (6.51)$$

The variance is thus proportional to the total flux through the image plane, less the flux which is specularly transmitted by the hologram. In other words, it is proportional to the total image-forming flux, or the total diffracted flux. Hence when we illuminate the hologram with irradiance R_o^2, we have

$$\text{Total diffracted flux} = 4\pi H R_o^2 V(t). \tag{6.52}$$

Carrying this analysis one step further, we can compare this light flux with that in a direct image of the object, produced as shown in Fig. 6.15c. This image is formed in the same position as the primary holographic image. For the direct image the amplitude in the hologram plane is given by $O_o(x)e^{i\varphi_o(x)}$. By Parseval's theorem,

$$\int_{-H}^{H} |O_o(x)e^{i\varphi_o(x)}|^2 \, dx = \int_{-H}^{H} |O_o|^2 \, dx = 2H\langle O_o^2 \rangle = \frac{1}{2\pi} \int_{-\infty}^{\infty} |S(\omega)|^2 d\omega, \tag{6.53}$$

where $S(\omega)$ is the Fourier transform of $O_o(x)e^{i\varphi_o(x)}$ and is the amplitude in the image plane. Therefore we can say

$$\text{Total flux in direct image} = 4\pi H \langle O_o^2 \rangle \tag{6.54}$$

and form the ratio

$$K \equiv \frac{\text{diffracted flux}}{\text{direct image flux}} = \frac{R_o^2}{\langle O_o^2 \rangle} V[t(x)] = \frac{V[t(x)]}{B} \tag{6.55}$$

where we have defined

$$B \equiv \frac{\langle O_o^2 \rangle}{R_o^2} \tag{6.56}$$

as the *beam balance ratio*. This is an important parameter in any hologram recording system and will be used often throughout this analysis. The quantity $\langle O_o^2 \rangle$ is just the average irradiance at the hologram plane due to the object and R_o^2 is the irradiance of the reference beam at the hologram plane.

It should be noted that while (6.55) has been derived for a Fourier transform hologram, this is not really a necessary requirement. For a Fresnel hologram, there is not an exact Fourier relationship between the object and hologram planes, nor between the hologram and image planes. There will always be a pair of planes (effectively at infinity), however, which have a Fourier transform relationship with the hologram plane. The near object of a Fresnel hologram may be considered to be an "out-of-focus" object with respect to the Fourier object plane and the image will similarly be out of focus with respect to the Fourier image plane. Since the ratio K is a ratio of light fluxes between these planes, it should be the same for either hologram type.

6.1.2.3   The Transfer Curve

In the preceding discussion, we made no assumptions about the recording medium except that a spatially varying exposure results in a spatially varying transmittance. In this section we will examine the role of the sensitometric characteristics of the recording medium (photographic emulsion) and calculate the variance of amplitude transmittance from the variance of exposure.

In pictorial photography it is convenient to show sensitometric data in terms of a density-log exposure curve such as shown in Fig. 6.16a. In holography, however, we are basically concerned with how the hologram diffracts light, which is determined by the variations of amplitude transmittance. This information can only be obtained indirectly from a D-log-E curve. A more convenient curve would then be an amplitude transmittance –exposure curve, or t–E, as shown in Fig. 6.16b. From an engineering point of view, however, this type of curve has the disadvantage that unless some

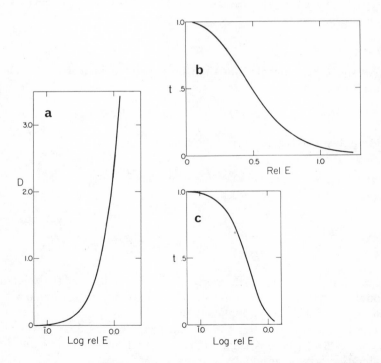

Fig. 6.16   Three possible transfer curves. (a) The conventional D-log E curve. (b) The t-E curve. (c) The t-log E curve.

sort of normalization is used, the shape depends on emulsion speed. Hence this curve would be inconvenient for comparing curves for different emulsions or processing. A third possible transfer curve which could be used is the t versus log-E curve such as shown in 6.16c. This curve has the advantage that its shape does not change with speed. Also, the variation in log exposure depends only on the ratio of irradiances of the two interfering beams and is independent of the average level. This is not so for the t versus E curve where doubling the exposure levels also doubles the exposure range. We expect that the flux diffracted from the hologram will correlate with the slope of curve 6.16b.

There is no correct curve which should be used—the choice should be based only on convenience. It is possible that the choice of transfer curve could depend on the application. In the following sections, each representation will be discussed in detail. Expressions will be derived for the variance of amplitude transmittance as a function of known parameters of the hologram.

A. AMPLITUDE TRANSMITTANCE VERSUS EXPOSURE CURVE. From (6.47) we have for the exposure

$$E(x) = E_o + F(x).$$

The t versus E relationship can then be expressed in terms of a shifted exposure coordinate as

$$t(x) = g[F(x)] = g[E(x) - E_o]. \tag{6.57}$$

Expanding this in a Taylor's series yields

$$t(x) = g(E_o) + g'(E_o)[E(x) - E_o] + g''(E_o)\frac{[E(x) - E_o]^2}{2!} + \cdots \tag{6.58}$$

where

$$g'(E_o) = \frac{dg}{dE}\bigg|_{E_o}, \qquad g''(E_o) = \frac{d^2g}{dE^2}\bigg|_{E_o}, \ldots$$

The expansion is in terms of the derivatives of g, that is, of the t versus E functional relationship. It is desirable that for the region of the curve that is of interest the derivatives higher than the first not be large, so that the series can be terminated after the second term. In any case, if the beam balance ratio B of (6.56) is small, $E(x) - E_o$ will be small and the linear approximation will be good.

Making this linear approximation, we can write

$$V[t(x)] = \langle t^2 \rangle - \langle t \rangle^2$$

$$= \langle \{g(E_o) + g'(E_o)[E(x) - E_o]\}^2 \rangle - \langle g(E_o) + g'(E_o)[E(x) - E_o] \rangle^2$$

$$= g^2 + 2gg'\langle E(x) - E_o \rangle + g'^2 \langle [E(x) - E_o]^2 \rangle$$

$$- g^2 - g'^2 \langle E(x) - E_o \rangle^2 - 2gg'\langle E(x) - E_o \rangle^2. \tag{6.59}$$

Now since $\langle E(x) \rangle \equiv E_o$, we have $\langle E(x) - E_o \rangle = 0$ and

$$V[t(x)] = g'^2(E_o)\langle [E(x) - E_o]^2 \rangle$$

$$= g'^2(E_o)[\langle E^2(x) \rangle - 2E_o\langle 2E(x) \rangle + E_o^2]$$

$$= g'^2(E_o)[\langle E^2(x) \rangle - E_o^2]$$

$$= [g'(E_o)]^2 V[E(x)]. \tag{6.60}$$

The exposure is given by (6.46) so that we can write

$$V[E(x)] = V[R_o^2] + V[O_o^2] + V\{2O_oR_o \cos [kx \sin \alpha_R - \varphi_o(x)]\}$$

$$+ \text{covariance terms}. \tag{6.61}$$

Both the variance of R_o^2 and the covariance between R_o^2 and any other term are zero since R_o^2 is a constant. The remaining covariance term is given by

$$CV\{O_o^2; O_oR_o \cos [kx \sin \alpha_R - \varphi_o(x)]\}$$

$$= \langle O_o^2 \cdot R_oO_o \cos [kx \sin \alpha_R - \varphi_o(x)] \rangle$$

$$- \langle O_o^2 \rangle \langle O_oR_o \cos [kx \sin \alpha_R - \varphi_o(x)] \rangle. \tag{6.62}$$

We now show that both terms on the right are zero. Writing

$$f_1(x) = O_o^2(x)$$

$$f_2(x) = O_oR_o \cos [kx \sin \alpha_R - \varphi_o(x)]$$

we have

$$\int_{-\infty}^{\infty} f_1(x)f_2^*(x)\, dx = \frac{1}{2\pi} \int_{-\infty}^{\infty} F_1(\omega)F_2(\omega)\, d\omega, \tag{6.63}$$

where $F_1(\omega)$ and $F_2(\omega)$ are the respective Fourier transforms of $f_1(x)$ and $f_2(x)$. Equation 6.63 is a generalized form of Parseval's theorem. Thus $F_1(\omega)$ refers to the distribution of flare light in the hologram plane, whereas $F_2(\omega)$ refers to the light distribution in the primary and conjugate images. If α_R is large enough so that these do not overlap, the right-hand side of (6.63) is zero since $F_1(\omega) = 0$ in the range of ω where $F_2(\omega) = 0$, and vice versa. Thus

$$\int_{-\infty}^{\infty} f_1(x) f_2^*(x)\, dx = 2H \langle f_1(x) f_2^*(x) \rangle = 0 \qquad (6.64)$$

since $f_1(x)$ and $f_2(x)$ are nonzero only for $|x| \geq H$, where $2H$ is the dimension of the hologram. Hence the first term on the right-hand side of (6.62) is zero.

For the second term, we write

$$f_2(x) = \frac{1}{2\pi} \int_{-\infty}^{\infty} F_2(\omega) e^{i\omega x}\, d\omega, \qquad (6.65)$$

so that

$$\langle f_2(x) \rangle = \frac{1}{4\pi H} \int_{-\infty}^{\infty} d\omega \int_{-H}^{H} F_2(\omega) e^{i\omega x}\, dx$$

$$= \frac{1}{2\pi} \int_{-\infty}^{\infty} F_2(\omega) \frac{\sin \omega H}{\omega \pi}\, d\omega = \frac{F_2(o)}{2H}, \qquad (6.66)$$

since $(\sin \omega H)/\pi\omega$ behaves like a δ-function for large H.

Since the spectrum of $f_2(x)$ contains no zero frequency, $\langle f_2(x) \rangle = 0$. Thus we have shown that both terms on the right of (6.62) are zero, so that

$$V[E(x)] = V[O_o^2] + V\{2O_o R_o \cos [kx \sin \alpha_R - \varphi_o(x)]\}. \qquad (6.67)$$

The first term on the right refers to the variance of the granular speckle pattern arising from the diffuse object beam. The variance of $O_o^2(x)$ can be determined by the same method Stone [13] uses for studying the Fraunhofer diffraction pattern of a screen containing a large number of small apertures of uniform size but randomly positioned. His analysis makes use of the fact that in the far field of the screen, the amplitude from each of the apertures will have the same modulus but a random phase. This same condition will be met for a simple diffuse object for which we can assume that the transmitted amplitude has a constant modulus but a random phase.

Stone's analysis makes use of the random walk process and will not be repeated here. The result, however, is that the standard deviation of the irradiance distribution in the far field is equal to the average irradiance. Since the variance is the square of the standard deviation, we have

$$V(O_o^2) = \langle O_o^2 \rangle^2. \qquad (6.68)$$

Also,

$$V\{2O_o R_o \cos [kx \sin \alpha_R - \varphi_o(x)]\}$$

$$= \langle 4R_o^2 O_o^2 \cos^2 (kx \sin \alpha_R - \varphi_o) \rangle - \langle 2R_o O_o \cos (kx \sin \alpha_R - \varphi_o) \rangle^2$$

$$= \langle 2O_o^2 R_o^2 [1 + \cos 2(kx \sin \alpha_R - \varphi_o)] \rangle - \langle 2O_o R_o \cos (kx \sin \alpha_R - \varphi_o) \rangle^2$$

$$= 2R_o^2 \langle O_o^2 \rangle + 2R_o^2 \langle O_o^2 \cos 2(kx \sin \alpha_R - \varphi_o) \rangle$$

$$\quad - \langle 2O_o R_o \cos (kx \sin \alpha_R - \varphi_o) \rangle^2. \qquad (6.69)$$

The second term on the right is zero for the same reason that $\langle f_2(x) \rangle = 0$ (Equation 6.66). The third term is $\langle f_2^2(x) \rangle$, therefore

$$V\{2O_oR_o \cos [kx \sin \alpha_R - \varphi_o(x)]\} = 2R_o^2 \langle O_o^2(x) \rangle. \qquad (6.70)$$

This leads to the result

$$V[E(x)] = \langle O_o^2 \rangle^2 + 2R_o^2 \langle O_o^2 \rangle. \qquad (6.71)$$

Since R_o^2 is the irradiance of the reference beam and $\langle O_o^2(x) \rangle$ that of the object beam, let us define

$$R_o^2 \equiv E_R$$

and

$$\langle O_o^2(x) \rangle \equiv E_S$$

so that

$$V[E(x)] = E_S^2 + 2E_R E_S. \qquad (6.72)$$

From (6.55), (6.56), and (6.60) we find that

$$K = \frac{E_S^2 + 2E_R E_S}{B} [g'(E_o)]^2. \qquad (6.73)$$

In terms of beam balance $B = E_S/E_R$ we have

$$K = [g'(E_o)]^2 E_R^2 (B + 2). \qquad (6.74)$$

The total average irradiance at the hologram plane is

$$E_o = E_R + E_S = E_R(1 + B)$$

therefore

$$K = [g'(E_o)]^2 \frac{E_o^2}{(1 + B)^2} (B + 2). \qquad (6.75)$$

Considering (6.72), we should recall that the first term on the right pertains to the flux diffracted by the speckle pattern produced by the object beam. This flux results in flare light around the illuminating beam, the angular extent of which is up to twice the field angle subtended by the object for plane holograms. The second term pertains to the flux diffracted into the primary and conjugate images. If these two components do not overlap, we can divide (6.75) into flare and image-forming components:

$$K_f = g'^2 \frac{BE_o^2}{(1 + B)^2} \qquad (6.76)$$

and

$$K_i = 2g'^2 \frac{E_o^2}{(1 + B)^2}. \qquad (6.77)$$

Both K_i and K_f can be calculated from the t versus E transfer curve of the recording medium if the beam balance ratio B is known. Thus knowledge of the transfer curve enables one to determine the flux diffracted into a holographic image relative to the flux in a direct image of the object.

B. AMPLITUDE TRANSMITTANCE VERSUS LOG-EXPOSURE CURVE. Using the same methods as for the t versus E curve, we can calculate t from the t versus log-E curve. Again we assume that the exposure (irradiance) at the hologram plane is a spatially varying function, varying now in $\log E$ about some average value $\langle \log E(x) \rangle$. Thus

$$t(x) = g[\log E(x) - \langle \log E(x) \rangle]$$
$$= g[\rho(x) - \rho_o] \tag{6.78}$$

where we have simplified the notation by defining

$$\rho(x) = \log E(x)$$
$$\rho_o = \langle \log E(x) \rangle. \tag{6.79}$$

Again we expand in a Taylor's series to give

$$t(x) = g(\rho_o) + g'(\rho_o)[\rho - \rho_o] + g''(\rho_o)\frac{[\rho - \rho_o]^2}{2!} + \cdots. \tag{6.80}$$

Terminating after the second term for the same reasons as before and using (6.46), we find

$$t(x) = g(\rho_o) - \rho_o g'(\rho_o) + g'(\rho_o)$$
$$\times \log \{R_o^2 + O_o^2(x) + 2R_o O_o(x) \cos [kx \sin \alpha_R - \varphi_o(x)]\}$$
$$= g(\rho_o) - \rho_o g'(\rho_o) + g'(\rho_o) \log R_o^2 + g'(\rho_o)$$
$$\times \log \{1 + G^2(x) + 2G(x) \cos [kx \sin \alpha_R - \varphi_o(x)]\} \tag{6.81}$$

where $G(x) = O_o(x)/R_o$. Collecting all constant terms, we obtain

$$t(x) = C + g'(\rho_o) \log \{1 + G^2(x) + 2G(x) \cos [kx \sin \alpha_R - \varphi_o(x)]\}. \tag{6.82}$$

For $|G(x)| < 1$, we can make the expansion

$$\log \{1 + G^2(x) + 2G(x) \cos [kx \sin \alpha_R - \varphi_o(x)]\}$$
$$= 2 \log_{10} e \left\{ \sum_{n=1}^{\infty} (-1)^{n+1} \frac{G^n(x)}{n} \cos n[kx \sin \alpha_R - \varphi_o(x)] \right\} \tag{6.83}$$

so that (6.82) becomes

$$t(x) = C + .868 g'(\rho_o) \left\{ \sum_{n=1}^{\infty} (-1)^{n+1} \frac{G^n(x)}{n} \cos n[kx \sin \alpha_R - \varphi_o(x)] \right\}. \tag{6.84}$$

As expected, the transmittance function contains a fundamental plus harmonics which produce higher-order diffraction. It is interesting to note that there is no longer a term containing $O_o(x)$ without a cosine factor. This implies that a plane hologram recorded in the linear region of the t versus log-E transfer curve would not produce the flare light commonly observed around the zero order. This has not been observed experimentally, however, possibly because the recording media used (photographic emulsions) were not of zero thickness or possibly because of the effects of the emulsion MTF, which have been ignored so far.

The variance of $t(x)$ can be obtained using (6.84) in a manner similar to that used to arrive at (6.67) and (6.71). The expression for K of (6.55) is concerned with the total amount of diffracted light in the far field, but we are concerned here with only the flux in the primary and conjugate first-order waves. Thus we want to extend the integration only over the spatial frequencies which include the first-order images. The complete variance of $t(x)$ obtained using (6.84) contains variance terms for the appropriate cosine terms for the various values of n, plus covariance terms. Since the first-order and the higher order images will not normally overlap (i.e., these covariances are zero), we need consider only the case for which $n = 1$. Thus we find that

$$V_1\{.868G(x) \cos [kx \sin \alpha_R - \varphi_o(x)]\} = .377B \qquad (6.85)$$

where B is the spatially averaged beam balance ratio defined by (6.56). Similarly,

$$V_1[t(x)] = .377g'^2(\rho_o)B \qquad (6.86)$$

where the subscript 1 indicates that the variance for $n = 1$ only is calculated. Finally, using (6.55), we find that

$$K = .377g'^2(\rho_o). \qquad (6.87)$$

Note that g in this equation is defined by (6.78) and is not the same g as that used in (6.77): in (6.87) g' refers to the gradient of the t versus log-E curve.

c. DENSITY VERSUS LOG-EXPOSURE CURVE. An analysis of the density versus log-exposure curve, or D log-E curve, will be given here for the sake of completeness and convenience, since virtually all of the published curves on commercially available photographic materials are of this type. For the transfer curve we can write

$$D = g(\log E). \qquad (6.88)$$

Then

$$t(x) = 10^{-D/2} = 10^{-g(\log E/2)} = h(\log E) = h(\rho) \qquad (6.89)$$

with $\rho = \log E$.

Again making a Taylor's expansion about the point of average log exposure $\rho_o \equiv \langle \log E \rangle$, we obtain

$$t(x) = h(\rho_o) + h'(\rho_o)(\rho - \rho_o) + h''(\rho_o)\frac{(\rho - \rho_o)^2}{2!} + \cdots. \qquad (6.90)$$

Since

$$h(\rho) = 10^{-g(\log E/2)}$$

and

$$h'(\rho) = \frac{dh}{d\rho} = \frac{dh}{dg}\frac{dg}{d\rho}, \qquad (6.91)$$

we have

$$h'(\rho) = -\frac{\log_e 10}{2}(10^{-g(\log E/2)})g'(\rho) = 1.15h(\rho)g'(\rho). \qquad (6.92)$$

Equation 6.90 then becomes

$$t(x) = h(\rho_o) + 1.15h(\rho_o)g'(\rho_o)(\rho - \rho_o) + \cdots. \qquad (6.93)$$

In order to terminate (6.90) at the first order, we must assume that $\rho - \rho_o$ is small, since even in the case of a linear D versus log-E curve the higher derivatives of h are not necessarily zero. In this case the variance becomes

$$V[t(x)] = 1.32h^2(\rho_o)[g'(\rho_o)]^2\langle(\rho - \rho_o)^2\rangle, \qquad (6.94)$$

where we have made use of the fact that $\langle \rho - \rho_o \rangle = 0$. Now

$$\langle(\rho - \rho_o)^2\rangle = \langle \rho^2 \rangle - \langle 2\rho\rho_o \rangle + \langle \rho_o \rangle^2 = \langle \rho^2 \rangle - \rho_o^2$$
$$= \langle \rho^2 \rangle - \langle \rho \rangle^2 = V(\rho) = V(\log E). \qquad (6.95)$$

Thus, since we are assuming small signals, (6.90) implies $h(\rho_o) = \langle t \rangle$, although it is not generally true that the function of an average value equals the average of the function. Based on the assumption of small signals, then,

$$V[t(x)] = 1.32\langle t \rangle^2 g'^2(\rho_o)V(\rho). \qquad (6.96)$$

The variance of $\log E$ corresponding to the image-producing part of the function describing the hologram transmittance is given by (6.85). Using this expression, we find

$$V_1[t(x)] = .497T[g'(\rho_o)]^2B \qquad (6.97)$$

where we have written $T \equiv \langle t \rangle^2$ for the average flux transmittance.

Thus we can see why the image radiance is greatest for exposure in the upper toe region of the D versus log-E curve: it is here that both the gradient (g') and the average flux transmittance (T) are relatively large.

Equation 6.97 has an interesting implication concerning the ratio of flux diffracted into the image to that specularly transmitted, that is, the ratio of first- to zero-order fluxes. From (6.52) we obtain

$$\text{total diffracted flux} = 4\pi H R_o^2 V[t(x)].$$

The total zero-order flux is given by $4\pi R_0^2 T$, so that

$$Q = \frac{\text{total diffracted flux}}{\text{zero-order flux}} = .497g'^2(\rho_o)B. \qquad (6.98)$$

Hence we expect that this ratio should remain constant for the straight-line portion of the D versus log-E curve.

6.1.2.4   *Experiments* [14]

A. INTRODUCTION.   A series of measurements has been made by Kaspar, Lamberts, and Edgett that indicates the general validity of the foregoing theory. In this section, their experimental results will be presented, showing how the amount of light flux diffracted into the primary image of a diffuser depends on the amplitude transmittance of the hologram, the beam balance ratio, and the object size. The holograms were all recorded on Kodak Spectroscopic Plates, Type 649F, with a mean spatial frequency of about 500 lines/mm.

B. AMPLITUDE   TRANSMITTANCE. Figure 6.17 shows the theoretical and measured ratio of the flux diffracted into the primary image to the flux in the direct image of the object. This ratio is denoted by $K_i/2$, where K_i is given by (6.77). Although there is some scatter in the experimental points, we see that there is a definite peak at an amplitude transmittance of about .5, or a specular optical density of about .6. The beam balance ratio is very small, about 1:90, ensuring a linear relationship between t and E. The actual t-E curve for these experiments is also shown in Fig. 6.17.

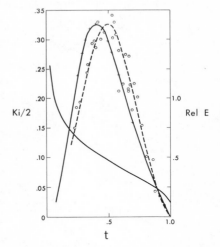

Fig. 6.17   Ratio of the flux diffracted into the primary image to the flux in the direct image as a function of amplitude transmittance t. The solid curve has been calculated from (6.75). The dashed curve is experimental. Superimposed on these curves is the actual t-E curve for these measurements [14].

The theoretical curve has been calculated from (6.75); corrections were made for a base and fog density of .10 and a modulation transfer factor for the emulsion of .76. The lateral displacement between the two curves has not yet been explained.

C. RATIO OF FIRST- TO ZERO-ORDER FLUX AS A FUNCTION OF DENSITY. Equation 6.98 indicates that Q, the first- to zero-order flux ratio, depends directly on the beam balance ratio during exposure and upon the square of the gradient of the D log-E curve (gamma). Thus if a density series of holograms is made, each at a common beam balance ratio, Q should be directly proportional to the squared gradient of the D log-E curve (γ^2) and should reach a maximum and become constant in the linear region.

Figure 6.18 shows an experimental curve of Q versus D and a curve of γ^2 versus D, with a proportionality constant chosen so that the two curves match at $D = 3$. Superposed on these two curves is the actual D versus log-E curve. The agreement is seen to be quite good.

D. BEAM-BALANCE RATIO. Of fundamental importance to anyone making holograms is the answer to the question, "What should be the relative strength of the object and reference beam?" Figure 6.19 shows experimentally determined values of $\frac{1}{2}K_i$ as a function of B (open circles). Figure

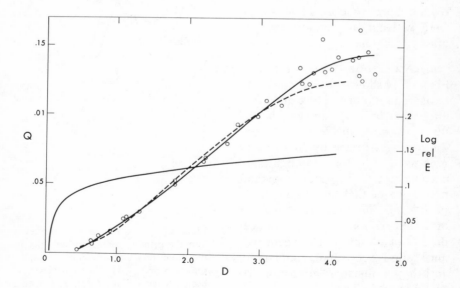

Fig. 6.18 The ratio of first- to zero-order flux (Q) as a function of density (solid curve) and also γ^2 as a function of density (dashed curve). Superimposed on these curves is the actual D-log E curve [14].

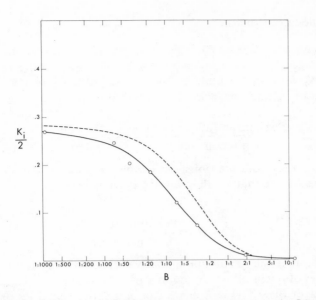

Fig. 6.19 The ratio of the flux diffracted into the primary image to the flux in the direct image as a function of the beam balance ratio B. The dashed curve has been calculated from Eq. 6.77 (see text) and the solid curve is experimental [14].

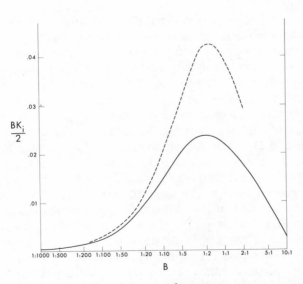

Fig. 6.20 The relative flux in the primary image $\frac{1}{2}K_i \cdot B$ as a function of B. The dashed curve has been calculated from Eq. 6.77 (see text) and the solid curve has been obtained from the solid curve of Fig. 6.19 [14].

6.20 shows a curve of $\frac{1}{2}(K_i \cdot B)$, which is proportional to the flux diffracted into the image. This curve indicates a peak of $B = \frac{1}{2}$ rather than 1. We say that $\frac{1}{2}K_i \cdot B$ is proportional to the flux diffracted into the image because, from the definitions of K_i and B, we have

$$\frac{1}{2}K_i \cdot B = \frac{\text{flux diffracted into one image}}{\text{direct image flux}} \cdot \frac{\text{object flux}}{\text{reference beam flux}}.$$

But since the flux from the object is identically equal to the direct image flux, and we assume that the illuminating beam is identical to the reference beam,

$$\frac{1}{2}K_i \cdot B = \frac{\text{flux diffracted}}{\text{flux incident}}.$$

By assuming that the maximum spatial variation in exposure is simply a standard deviation in either direction from the mean exposure, and that the gradient of the t versus E curve can be approximated by the average gradient between these resulting maximum and minimum exposures, we can use (6.77) to obtain the dashed curves of Figs. 6.19 and 6.20. The standard deviation used here contains both terms of (6.72), that is, it includes the contributions from the speckle pattern which results from the diffuse nature of the object.

It is interesting to note that the theoretical curve (using the previously mentioned approximations) for relative diffracted flux ($\frac{1}{2}K_i \cdot B$) also peaks at a beam balance ratio of about $\frac{1}{2}$ rather than 1. This apparently is caused by sensitometric nonlinearity when adding the effect of the speckle pattern to the image-producing transmittance term on the hologram. For large beam balance ratios (approaching 1 and greater) the speckle is relatively more pronounced and produces a large amount of flare light around the zero order, and this flux does not contribute to the image. This effect is shown in Fig. 6.21. These are photomicrographs of actual holograms recorded with decreasing values of B. In (a) $B = 10$, (b) 5, (c) 2, (d) 1, (e) $\frac{1}{3}$, and (f) $\frac{1}{10}$. The image-forming flux is diffracted by the very fine fringes, not by the coarser speckle. For large B, these fringes are swamped by the speckle, reducing the amount of flux diffracted into the images.

E. OBJECT SIZE.   The light flux diffracted into the image depends directly on the variance of the amplitude transmittance of the hologram. In the case of a linear t versus E relationship, the variance depends only on the slope of the t versus E curve, the average exposure, and the beam balance ratio. When the relation between t and log E is linear, the variance will depend only on the slope and beam balance ratio. In neither case does the geometry of the object affect the amount of flux diffracted, unless, of course,

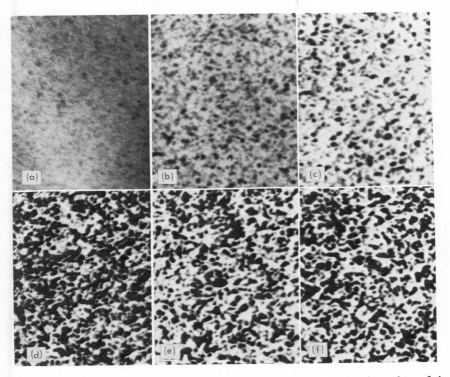

Fig. 6.21 Photomicrographs of actual holograms recorded with decreasing values of the beam balance ratio B. In (a) $B = 10$, (b) 5, (c) 2, (d) 1, (e) $\frac{1}{3}$, and (f) $\frac{1}{10}$.

the geometry influences the beam balance ratio. This is probably true whether or not linearity exists since, for example, in the case of the t versus log-E curve, the variance of t depends only on the derivatives of the sensitometric curve and the beam balance ratio regardless of the nonlinearity between t and log E [see, e.g., (6.78) et seq.].

It follows from this assumption that if we make a series of holograms with increasing object size while maintaining a constant beam balance ratio, the flux in each of the images should be the same. Thus for uniformly irradiated diffuse objects, the product of image area and irradiance should be the same for all holograms (as it is for the objects themselves when the beam balance is kept constant).

Figure 6.22 shows the results of measurements made on several holograms made with diffusers of various sizes and $B = 1:7.5$. It is seen that $K_i/2$ remains constant as the object size increases, as it should. The irradi-

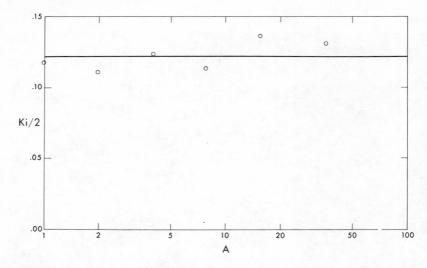

Fig. 6.22 The ratio of the flux diffracted into the primary image to the flux in the direct image as a function of the object area A [14].

ance of the image and object are decreasing with increasing object size, since the beam balance ratio is maintained at $1:7.5$.

6.2   OBJECT MOTION

6.2.1   Introduction

Since forming the hologram record is essentially a two-beam interference experiment, it is important that the phase difference between the two beams remain constant during the exposure. A change in the phase difference at the recording plane of π will shift the fringe position by $\frac{1}{2}$ of one fringe spacing. If this shift occurs during the exposure, the recorded fringe modulation will fall to practically zero and no hologram will be recorded. Time-varying phase differences at the hologram plane can occur in many ways. One of these, a time-varying wavelength from the source, was discussed in Section 6.0. The phase difference between the object and reference beams can also change if one of the beam-directing mirrors moves or if the object itself moves. Although the title of this section is "Object Motion," the results can be applied to any cause of a changing phase difference at the hologram. In general, if the object vibrates or moves during the exposure, the change in phase difference at the hologram will not be the same at all points of the hologram. The complete description of the exposure recorded in the presence of object motion will depend on the geometrical arrangement used and the type of hologram being recorded.

In general, we can say that a hologram of a transparency or back-lighted scene is less sensitive to object motion than is a hologram of an opaque, reflecting object. Motion of the recording medium itself during an exposure time can be no more than a small fraction of the fringe spacing. Many of the effects of motion of the recording medium are similar to those of object motion.

These effects can be analyzed in several ways, but one of the most generally applicable is that used by Goodman [15] in which the hologram is regarded as a temporal filter. This, then, will be our approach.

6.2.2 The Hologram as a Temporal Filter

In order to consider the problem of object motion during recording, we write the exposure at the point x of the recording medium as

$$E(x) = \int_{-T/2}^{T/2} | H(x, t)|^2 \, dt \tag{6.99}$$

where, as before, $| H |^2$ is the irradiance distribution at the recording medium (now a function of time) and T is the exposure time. We can write H as

$$H(x, t) = R(x) + O(x, t) \tag{6.100}$$

so that the exposure is

$$E(x) = E_R + E_S + R(x)\int_{-T/2}^{T/2} O^*(x, t) \, dt + R^*(x)\int_{-T/2}^{T/2} O(x, t) \, dt \tag{6.101}$$

where E_R and E_S are the exposures due to the reference and object waves alone, respectively. We will restrict our attention to the last term which leads to the primary image. Calling this term E_p, we obtain

$$E_p = R^*(x)\int_{-T/2}^{T/2} O(x, t) \, dt \tag{6.102}$$

which can be rewritten as

$$E_p = R^*(x)\int_{-\infty}^{\infty} \text{rect}\left(\frac{t}{T}\right) O(x, t) \, dt \tag{6.103}$$

where

$$\text{rect}\left(\frac{t}{T}\right) = \begin{cases} 1 & -\dfrac{T}{2} \le t \le \dfrac{T}{2} \\ 0 & \text{otherwise} \end{cases} \tag{6.104}$$

If $\tilde{O}(x, \nu)$ is the Fourier transform of $O(x, t)$, defined by

$$\tilde{O}(x, \nu) = \int_{-\infty}^{\infty} O(x, t)e^{-2\pi i \nu t}\, dt \qquad (6.105)$$

we may use Parseval's theorem [16] to write (6.103) as

$$E_p = TR^*(x)\int_{-\infty}^{\infty} \text{sinc}\,(\pi \nu T)\, \tilde{O}(x, \nu)\, d\nu \qquad (6.106)$$

where $T\,\text{sinc}\,[\pi \nu T] = \sin\,[\pi \nu T]/\pi \nu$ and is the Fourier transform of rect (t/T). The interpretation given (6.106) is that of a linear filter having a transfer function $\text{sinc}\,(\pi \nu T)$. If the frequencies of the reference and object beams differ by ν, the amplitude of the field at the point x will be reduced by a factor $\text{sinc}\,(\pi \nu T)$.

6.2.3   Linear Motion of the Object [17]

A simple application of (6.106) can be made where an object point is moving with a constant velocity v in a straight line. Assume that we have the basic recording configuration of Fig. 6.23. The point B represents the beam-splitter where the object and reference beams divide. The point R is the source of the reference beam. The point $S(x_o, y_o)$ is a typical object point which we assume to be moving with a velocity $v = |V|$ in the direction of the vector V defined by the two angles α and β (we assume that the distance moved by this object point S during T is so small that α and β remain constant). The phase difference at the point $Q(x, z)$ in the recording medium is given by

$$\delta(Q) = k(\overline{BRQ} - \overline{BSQ}), \qquad (6.107)$$

where $k = 2\pi/\lambda$. Let us call $k\,\overline{BRQ} = \varphi_R$ and $k\,\overline{BSQ} = \varphi(t)$. The optical path $\overline{BS}$ is increasing at a rate $-|V|\cos \alpha = -v\cos \alpha$ and the path $\overline{SQ}$ is decreasing at a rate $v\cos \beta$. Hence the rate at which $\varphi(t)$ is decreasing is just

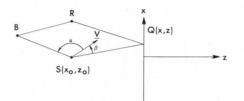

Fig. 6.23   Notation and recording arrangement for describing the effects of object motion.

$$\frac{d\varphi}{dt} = kv(\cos\alpha + \cos\beta) \tag{6.108}$$

so that

$$Q(t) = \varphi_o + kvt(\cos\alpha + \cos\beta) \tag{6.109}$$

where φ_o is simply the phase of the object wave at Q midway through the exposure ($t = 0$). The phase difference at Q is now written

$$\delta(Q) = \varphi_R - \varphi_o - kvt(\cos\alpha + \cos\beta). \tag{6.110}$$

This time-dependent phase difference results in fringe motion at Q during recording, with a subsequent loss of fringe modulation or contrast at Q.

To find the exposure at Q, we return to (6.106). Using (6.109), we can write

$$O(x, t) = O_o(x)e^{i\varphi(t)} = O_o(x)e^{i\varphi_o(x)}e^{ikvt(\cos\alpha + \cos\beta)}. \tag{6.111}$$

From (6.105),

$$\tilde{O}(x, v) = O_o e^{i\varphi_o} \int_{-\infty}^{\infty} e^{it[kv(\cos\alpha + \cos\beta) - 2\pi v]} \, dt$$

$$= O_o e^{i\varphi_o} \delta\left[\frac{v}{\lambda}(\cos\alpha + \cos\beta) - v\right] \tag{6.112}$$

where we have used the relations

$$\delta(x) = \frac{1}{2\pi} \int_{-\infty}^{\infty} e^{ikx} \, dk$$

and $\tag{6.113}$

$$\delta(ax) = \frac{1}{a}\delta(x)$$

for the Dirac delta function $\delta(x)$. Using the well-known sifting property of this function, (6.106) yields

$$E_p = R^*(x)O_o(x)e^{i\varphi_o(x)} \cdot T \operatorname{sinc}\left[\frac{kvT}{2}(\cos\alpha + \cos\beta)\right]. \tag{6.114}$$

The presence of the sinc function in this expression means that the recorded fringe modulation will not be constant over the hologram. This can easily be seen if we write the fringe modulation as

$$M = \frac{2E_p E_c}{E_S + E_R} \tag{6.115}$$

where E_S and E_R have been defined following (6.101) and E_p in (6.102). The exposure E_c is just that part of the exposure leading to the conjugate

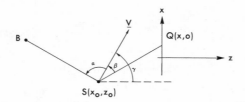

Fig. 6.24 Defining the angle γ between the velocity vector **V** and the z-axis.

image. The way in which M varies over the hologram can best be seen if we write (6.114) in a slightly different way. First we define a new angle γ (Fig. 6.24)—this is just the angle the velocity vector of the object point makes with the z-axis. Then we can write

$$\cos \beta = \cos [\gamma - (\gamma - \beta)] = \cos \gamma \cos (\gamma - \beta) + \sin \gamma \sin (\gamma - \beta). \quad (6.116)$$

If the angle $\gamma\text{-}\beta$ is small, we can write

$$\cos (\gamma - \beta) \approx \left[1 - \frac{(x - x_o)^2}{2z_o^2} \right]$$

$$\sin (\gamma - \beta) \approx \frac{x - x_o}{z_o}. \quad (6.117)$$

Defining the argument of the sinc function in (6.114) as Γ, we can write

$$\Gamma = \frac{kvT}{2} (\cos \alpha + \cos \beta)$$

$$= \frac{kvT}{2} \left[\cos \alpha + \left(1 - \frac{(x - x_o)^2}{2z_o^2} \right) \cos \gamma + \frac{(x - x_o)}{z_o} \sin \gamma \right]. \quad (6.118)$$

The quantity $(x - x_o)/z_o$ is just the half-angle subtended by the hologram out to the point x at the object point S. It is very closely related to the effective numerical aperture of the hologram, and for $x_o = 0$, it *is* the effective numerical aperture of the hologram. Since the numerical aperture of the hologram determines the diffraction-limited image resolution, we see that $(x - x_o)/z_o$ is an important parameter. We shall determine the maximum possible value for the quantity $(x - x_o)/z_o$ for several particular cases of object motion.

The argument Γ is zero for all $(x - x_o)/z_o$ only for the case of no motion, $v = 0$. In this case M will be a maximum over the hologram. For nonzero values of v, the first zero of E_p (hence M) occurs for $\Gamma = \pm\pi$. For a given choice of α, γ, and v, this value of Γ yields the maximum value for $(x - x_o)/z_o$. We will call a the number of wavelengths the object point S moves during the exposure time T: $vT = a\lambda$. Figure 6.25 shows curves of $[(x - x_o)/z_o]_{max}$ versus a for several typical holographic situations. Case I is for object illumination in the same direction, that is, $\alpha = \pi$, $\beta = 0$, $\gamma = 0$. For this case $\Gamma = -\pi$. It is seen that this configuration yields the largest effective hologram size. Case II is for $\alpha = \pi/2$, $\beta = \pi/2$, and $\gamma = \pi/2$. In this case the object is moving parallel to the hologram (in the x-direction) and the object illumination is along the z-axis. For Case III, $\alpha = \pi/2$ and $\beta = \gamma = 0$, that is, motion toward the hologram and illumination parallel to the hologram. Case IV corresponds to a configuration

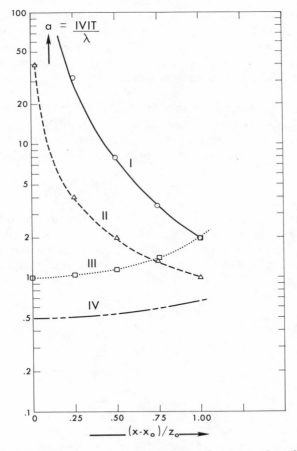

Fig. 6.25 The parameter $[(x-x_o/z_o)]_{max}$, as a function of a, the number of wavelengths of object motion, for four common geometries. (See text for an explanation of each of the curves.)

for recording a hologram of a diffusely reflecting object. Here $\alpha = \beta = \gamma = 0$; the motion is along the z-axis. This situation yields the smallest possible value for the effective hologram size, or alternatively, for recording a given-size hologram, this configuration is the most sensitive to object motion.

6.2.4 Effect of Linear Object Motion on the Image [17]

One way to determine what sort of image is formed with a hologram recorded in the presence of object motion is to guess at some hypothetical object which would, if illuminated as the real object was during recording, yield a particular field distribution in the hologram plane. This desired field distribution is, of course, just the wave (corresponding to the primary image) which is transmitted by the hologram when the hologram is illuminated with a wave identical to the original reference wave. The image formed with this wave will then be identical to an image of the hypothetical object.

We make the assumption that the final amplitude transmittance of the hologram is simply equal to the exposure. Then if the hologram is illuminated with a wave identical to the original reference wave, the transmitted field $\psi_p(x)$ which leads to the primary image is just $[R(x)]E_p$. Using (6.114), this field becomes

$$\psi_p(x) = |R(x)|^2 O_o(x)e^{i\varphi_o(x)}T \operatorname{sinc}\Gamma. \qquad (6.119)$$

We must now determine a source that will give rise to $\psi_p(x)$ at the hologram.

As in Fig. 6.26, assume that we have a line source of length a lying in the direction of the velocity v. We let l be the coordinate along the line source and assume a linear phase shift kbl along the source. From (3.2), the field at the point Q is

$$\psi(Q) = C\int_{-a/2}^{a/2} \frac{e^{ikr}}{r} e^{ikbl}\, dl \qquad (6.120)$$

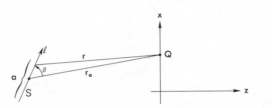

Fig. 6.26 A hypothetical source which will yield the field described by Eq. 6.119 in the x-plane.

where the constant C includes the obliquity and phase factors of (3.2) and we are here considering only a two-dimensional problem. We approximate the distance r in the denominator as r_o and write $r = r_o - l \cos \beta$ so that (6.120) becomes

$$\psi(Q) = C \frac{e^{ikr_o}}{r_o} \int_{-a/2}^{a/2} e^{ikl(b-\cos \beta)} \, dl$$

$$= C \frac{e^{ikr_o}}{r_o} a \operatorname{sinc} \left[\frac{ka}{2} (b - \cos \beta) \right]. \tag{6.121}$$

Equating Γ of (6.118) with $(ka/2)(b - \cos \beta)$, we find that the proposed source gives rise to a field distribution at the hologram which is identical to the transmitted primary image wave $\psi_p(x)$ if

$$a = vT$$
$$b = -\cos \alpha. \tag{6.122}$$

Thus the primary image of the moving point S will be a line image of length vT. The same arguments apply to the conjugate image. Hence the images formed with holograms which have been recorded in the presence of linear object motion should appear unsharp to the same degree as a conventional photograph of a moving scene, with the same motion (which we have assumed to be small) and exposure time.

6.2.5 Oscillatory Motion of the Object

The type of object motion most often encountered in practice is probably that caused by vibration—either of the object itself or of any of the optical elements. We need not restrict ourselves to object motion alone in this section, since a mirror or lens vibration during exposure will often be equivalent to object motion. Hence we will consider the problem of a time-dependent phase *difference* between the object and reference beams at the hologram plane. For the purpose of analysis, we will consider the simple but basic and not necessarily unrealistic case of a sinusoidal time dependence for the phase of the object beam. We assume the form to be

$$\varphi(t) = \varphi_o + ka \sin \omega t \tag{6.123}$$

analogous to (6.109). Hence

$$O(x, t) = O_o(x) e^{i \varphi(t)} = O_o(x) e^{i \varphi_o} e^{ika \sin \omega t} \tag{6.124}$$

and from (6.105)

$$\tilde{O}(x, \nu) = O_o e^{i\varphi_o} \int_{-\infty}^{\infty} e^{ika \sin \omega t} e^{-2\pi i\nu t} \, dt$$

$$= O_o e^{i\varphi_o} \int_{-\infty}^{\infty} \cos(ka \sin \omega t) e^{-2\pi i\nu t} \, dt$$

$$+ iO_o e^{i\varphi_o} \int_{-\infty}^{\infty} \sin(ka \sin \omega t) e^{-2\pi i\nu t} \, dt$$

$$= O_o e^{i\varphi_o} \left\{ J_o(ka) \int_{-\infty}^{\infty} e^{-2\pi i\nu t} \, dt \right.$$

$$+ 2 \sum_{n=1}^{\infty} \int_{-\infty}^{\infty} J_{2n}(ka) \cos(2n\omega t) e^{-2\pi i\nu t} \, dt$$

$$\left. + 2 \sum_{n=0}^{\infty} \int_{-\infty}^{\infty} J_{2n+1}(ka) \sin[(2n+1)\omega t] e^{-2\pi i\nu t} \, dt \right\} \tag{6.125}$$

where we have used the expansions

$$\cos(ka \sin \omega t) = J_o(ka) + 2 \sum_{n=1}^{\infty} J_{2n}(ka) \sin 2n\omega t$$

and $\tag{6.126}$

$$\sin(ka \sin \omega t) = 2 \sum_{n=0}^{\infty} J_{2n+1}(ka) \sin[(2n+1)\omega t].$$

The $J_n(ka)$ are Bessel functions of the first kind. Evaluating the integrals of (6.125) and rearranging the summations, we obtain

$$\tilde{O}(x, \nu) = O_o e^{i\varphi_o} \sum_{n=-\infty}^{\infty} J_n(ka) \, \delta\left(\nu - \frac{n\omega}{2\pi}\right). \tag{6.127}$$

Substitution of this expression into (6.106) yields

$$E_p = R^* T O_o e^{i\varphi_o} \sum_{n=-\infty}^{\infty} \left[J_n(ka) \int_{-\infty}^{\infty} \text{sinc}(\pi \nu T) \, \delta\left(\nu - \frac{n\omega}{2\pi}\right) d\nu \right]$$

$$= T R^* O_o e^{i\varphi_o} \sum_{n=-\infty}^{\infty} J_n(ka) \, \text{sinc}\left(\frac{n\omega T}{2}\right). \tag{6.128}$$

If ωT is large (say > 10), then $\text{sinc}(n\omega T/2)$ is small for all n and the major contribution to E_p comes from the $n = 0$ term. The quantity ωT is just the number of cycles completed by the phase (6.123) during an exposure time. Since a typical vibration frequency is of the order of a few cycles per second, and exposures are typically several seconds long, the approximation $\omega T > 10$ will usually be satisfied. With this approximation, then,

$$E_p = T R^* O_o e^{i\varphi_o} J_o(ka). \tag{6.129}$$

To obtain a more specific expression, we assume that the object is vibrating sinusoidally in a direction $\mathbf{V}$ as in Fig. 6.23. If the instantaneous speed of the object point S in the direction $\mathbf{V}$ is $|\mathbf{V}|$, then

$$\frac{d\varphi}{dt} = k|\mathbf{V}|(\cos\alpha + \cos\beta) \qquad (6.130)$$

as in (6.108). Since we are assuming a sinusoidal motion, we have for the instantaneous speed

$$|\mathbf{V}| = \frac{d}{dt}(a\sin\omega t) = a\omega\cos\omega t, \qquad (6.131)$$

so that

$$\frac{d\varphi}{dt} = ka\omega(\cos\alpha + \cos\beta)\cos\omega t \qquad (6.132)$$

and

$$\varphi(t) = \varphi_o + ka(\cos\alpha + \cos\beta)\sin\omega t. \qquad (6.133)$$

Hence the Bessel function argument ka of (6.129) should be replaced with $ka(\cos\alpha + \cos\beta)$ for the case of object motion—(6.129) applies only for the general phase variation of (6.123). For object motion, we have

$$E_p = TR^*O_o e^{i\varphi_o}J_0[ka(\cos\alpha + \cos\beta)]. \qquad (6.134)$$

In this case the fringe contrast is modulated by the zero-order Bessel function, whereas for the case of linear object motion, the modulating function was a sinc function. The same general statements about the various directions of motion and illumination made in reference to linear motion also apply here. The curves corresponding to Fig. 6.25 would be quite similar in appearance.

6.2.6 Effect of Oscillatory Object Motion on the Image

To see what the primary image of a vibrating object point looks like, we again find a source which yields the same field distribution at the hologram as the primary image wave. The latter is formed by illuminating the hologram of (6.134) with the reference wave R. The transmitted field is then

$$\psi_p(x) = R(x)\,{}^2O_o(x)e^{i\varphi_o(x)}\cdot T\cdot J_0[ka(\cos\alpha + \cos\beta)]. \qquad (6.135)$$

A hypothetical source giving rise to this field distribution is quite similar to the one found for linear object motion. We again assume a line source such as shown in Fig. 6.26, with a linear phase shift along l of kbl as before.

The length now, however, is taken to be $2a$ instead of a as shown in the figure. We assume that the amplitude is distributed along l as $[1 - (l/a)]^{2-\frac{1}{2}}$ so that the field at Q is

$$\psi(Q) = C \int_{-a}^{a} \left[1 - \left(\frac{l}{a}\right)^{2}\right]^{-\frac{1}{2}} \frac{e^{ikr}}{r} e^{ikbl} \, dl \qquad (6.136)$$

analogous to (6.120). Making the same approximations which led to (6.121), and writing

$$\frac{l}{a} = \sin x$$

and   (6.137)

$$dl = a \cos x \, dx,$$

we obtain

$$\psi(Q) = \frac{Cae^{ikr_o}}{r_o} \int_{-\pi/2}^{\pi/2} e^{ika \sin x(b-\cos \beta)} \, dx$$

$$= \frac{Cae^{ikr_o}}{r_o} \int_{-\pi/2}^{\pi/2} \cos\left[ka \sin x(b - \cos \beta)\right] \, dx$$

$$= \frac{Cae^{ikr_o}}{r_o} \int_{-\pi/2}^{\pi/2} \left\{ J_o[ka(b - \cos\beta)] + 2 \sum_{n=1}^{\infty} J_{2n}[ka(b - \cos\beta)] \sin 2nx \right\} dx$$

$$= \frac{\pi Cae^{ikr_o}}{r_o} J_o[ka(b - \cos \beta)]. \qquad (6.138)$$

Except for the constants, we see that this is of the same form as (6.135) if we make the identification $b = \cos \alpha$ as before. Hence we can say that the primary image of an oscillating object point is a line of length $2a$ and an amplitude which is proportional to $[1 - (l/a)^2]^{-\frac{1}{2}}$. The fact that the amplitude is infinite at the end points is unfortunate since this obviously cannot exist physically. We must consider this hypothetical source to be in the same realm as a point source—just a mathematical convenience. The fact that the end points of the line image will appear brighter is expected, since the oscillating point is stationary at the end points. Since a hologram of a moving point can be considered to be an incoherent superposition of many holograms, one for each position of the point, most of the flux in the resulting image will pass through the image points corresponding to the end points of the motion.

6.2.7   Effect of Motion of the Recording Medium

The problem of motion of the recording medium during exposure will be treated only for the case of lateral motion perpendicular to the direction of the fringes. Longitudinal motion will have only a second-order effect on the recorded fringes and, except in the case of very large motions, can be ignored. Lateral motion perpendicular to the fringes, on the other hand, will tend to wash out or demodulate the recorded fringes.

In order to examine this problem, we must backtrack to (6.102). With the recording medium itself in motion, the phase of the reference wave at the point x will be time-dependent and so must be included in the integral. Hence we write

$$E_p(x) = \int_{-T/2}^{T/2} R^*(x, t)O(x, t)\, dt \tag{6.139}$$

which we again write as

$$E_p(x) = \int_{-\infty}^{\infty} \text{rect}\left(\frac{t}{T}\right) R^*(x, t)O(x, t)\, dt. \tag{6.140}$$

By defining $\mathfrak{F}(\nu)$ as the Fourier transform of $R^*(x, t)O(x, t)$,

$$\mathfrak{F}(\nu) \equiv \widehat{R^*(x, \nu)O(x, \nu)} = \int_{-\infty}^{\infty} R^*(x, t)O(x, t)e^{-2\pi i\nu t}\, dt, \tag{6.141}$$

we can write

$$E_p(x) = T \int_{-\infty}^{\infty} \text{sinc}\, (\pi\nu T)\mathfrak{F}(\nu)\, d\nu \tag{6.142}$$

analogous to (6.106).

Consider Fig. 6.27. The object point is at S and the line joining S to a point Q of the hologram makes an angle β with the z-axis. (Note that this is a different definition of β than we had previously.) The reference beam R makes an angle α_R with the z-axis and the recording medium is assumed to have a constant velocity $\mathbf{V}$ in the x-direction. The phase of the reference wave at Q will change with time according to

$$\varphi_R(t) = \varphi_{R_o} + k|\mathbf{V}|\, t \sin \alpha_R \tag{6.143}$$

with a similar expression for the phase of the object wave at Q:

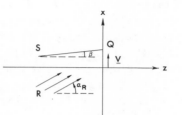

Fig. 6.27 Geometry and notation for discussing the effects of motion of the recording medium.

$$\varphi(t) = \varphi_o + k|\mathbf{V}|t\sin\beta. \tag{6.144}$$

Now

$$R^*(x, t)O(x, t) = R_oO_oe^{i(\varphi_o-\varphi_{R_o})}e^{ik|\mathbf{V}|t(\sin\beta-\sin\alpha_R)} \tag{6.145}$$

so that

$$\mathfrak{F}(\nu) = R_oO_oe^{i(\varphi_o-\varphi_{R_o})}\int_{-\infty}^{\infty}e^{ik|\mathbf{V}|t(\sin\beta-\sin\alpha_R)}e^{-2\pi i\nu t}\,dt$$

$$= R_oO_oe^{i(\varphi_o-\varphi_{R_o})}\,\delta\left[\nu - \frac{k|\mathbf{V}|(\sin\beta - \sin\alpha_R)}{2\pi}\right]. \tag{6.146}$$

Substituting (6.146) into (6.142), we obtain

$$E_p(x) = TR_oO_oe^{i(\varphi_o-\varphi_{R_o})}\int_{-\infty}^{\infty}\mathrm{sinc}\,(\pi\nu T)\delta\left[\nu - \frac{k|\mathbf{V}|(\sin\beta - \sin\alpha_R)}{2\pi}\right]d\nu$$

$$= TR_oO_oe^{i(\varphi_o-\varphi_{R_o})}\,\mathrm{sinc}\left[\frac{k|\mathbf{V}|T}{2}(\sin\beta - \sin\alpha_R)\right]. \tag{6.147}$$

Again, as in (6.114), the presence of the sinc function means that the fringe modulation varies over the hologram. Since the sinc function has zeroes whenever its argument is an integral multiple of π, we see that points on the hologram for which

$$\sin\beta - \sin\alpha_R = \frac{\lambda}{|\mathbf{V}|T} \tag{6.148}$$

have zero fringe contrast. But the quantity

$$\frac{\sin\beta - \sin\alpha_R}{\lambda} = \nu_s \tag{6.149}$$

is just the spatial frequency of the fringes at Q. This leads to a simple interpretation of (6.147). In a Fourier transform system, each point of the object corresponds to a different spatial frequency at the hologram plane. In this case, motion of the recording medium will lead to the disappearance of various spatial frequencies. Since these correspond to different object points, we expect that the resulting image formed with the hologram will contain dark fringes through the image field corresponding to the missing frequencies. We would expect that the results would not be too different for a Fresnel hologram, even though in the case of a Fresnel hologram, the spatial frequency for each point of the object varies over the hologram and hence, in general, will not be completely lost. It is likely, however, that *most* of the information about a certain region of the object will be lost, resulting in fringes in the image.

The case of oscillatory motion of the recording medium is not too different. We begin by writing the phase difference at Q as

$$\delta(x, t) = \varphi(t) - \varphi_R(t) = \delta_0 + kv(\sin \beta - \sin \alpha_R) \sin \omega t \quad (6.150)$$

where δ_o is the phase difference at $t = 0$ and v is the modulus of the velocity vector. We now find, similar to (6.145),

$$R^*(x, t)O(x, t) = R_o O_o e^{i\delta_o} \exp [ikv(\sin \beta - \sin \alpha_R) \sin \omega t] \quad (6.151)$$

so that

$$\mathfrak{F}(\nu) = R_o O_o e^{i\delta_o} \int_{-\infty}^{\infty} \exp [ikv(\sin \beta - \sin \alpha_R) \sin \omega t] e^{-2\pi i \nu t} \, dt. \quad (6.152)$$

Since this is now of the same form as (6.125), we can write immediately

$$\mathfrak{F}(\nu) = R_o O_o e^{i\delta_o} \sum_{n=-\infty}^{\infty} J_n[kv(\sin \beta - \sin \alpha_R)] \, \delta \left(\nu - \frac{n\omega}{2\pi} \right). \quad (6.153)$$

Substitution of this expression into (6.142) yields

$$E_p(x) = TR_o O_o e^{i\delta_o} \sum_{n=-\infty}^{\infty} J_n[kv(\sin \beta - \sin \alpha_R)] \, \text{sinc} \left(\frac{n\omega T}{2} \right). \quad (6.154)$$

Again, if $\omega T > 10$, for example, we can neglect the higher order Bessel functions and write

$$E_p(x) = TR_o O_o e^{i\delta_o} J_o[kv(\sin \beta - \sin \alpha_R)]. \quad (6.155)$$

The image formed with this hologram will be essentially the same as in the case of linear motion; there are some spatial frequencies (object points) that do not get recorded, hence there will be fringes in the image.

REFERENCES

[1] A. L. Bloom, *Appl. Opt.*, **5**, 1500 (1966).
[2] M. Born and E. Wolf, *Principles of Optics*, Pergamon Press Ltd., London, 1959, p. 319.
[3] M. Hercher, *Appl. Phys. Letters*, **7**, 39 (1965).
[4] F. J. McClung and D. Wiener, *IEEE J. Quantum Electronics*, QE-1, 94 (1965).
[5] A. D. Jacobson and F. J. McClung, *Appl. Opt.*, **4**, *1509* (1965).
[6] R. E. Brooks, L. O. Heflinger, and R. F. Wuerker, *IEEE J. Quantum Electronics*, QE-2, 275 (1966).
[7] R. E. Brooks, R. F. Wuerker, L. O. Heflinger, and C. Knox, Paper presented at the International Colloquium on Gasdynamics of Explosions, Brussels, Belgium, September 20, 1967.
[8] J. M. Burch, J. W. Gates, R. G. N. Hill, and L. H. Tanner, *Nature* **212**, 1347 (1966).
[9] R. W. Ditchburn, *Light*, Blackie and Son, Ltd., London, 1953, pp. 145–148.
[10] E. M. Leith and J. Upatnieks, *J. Opt. Soc. Am.*, **57**, 975 (1967).

[11] E. N. Leith, The Hologram Technique and Potential Applications, Radar Laboratory, Institute of Science and Technology, University of Michigan. Contract No. AF 18(600)-2779. AD No. 469650 (June 1965).

[12] F. G. Kaspar and R. L. Lamberts, *J. Opt. Soc. Am.*, **58,** 970 (1968).

[13] J. M. Stone, *Radiation and Optics*, McGraw-Hill, New York, 1963, p. 146.

[14] F. G. Kaspar, R. L. Lamberts, and C. D. Edgett, *J. Opt. Soc. Am.*, **58** (1968).

[15] J. W. Goodman, *Appl. Opt.* **6,** 857 (1967).

[16] A. Papoulis, *The Fourier Integral and Its Applications*, McGraw-Hill, New York, 1963, p. 27.

[17] D. B. Neumann, *J. Opt. Soc. Am.*, **58,** 447 (1968).

r

7 Color Holography

7.0 INTRODUCTION

Although so far we have discussed only holograms recorded in monochromatic light and have not as yet discussed the possibility of superposing more than one hologram in a single recording medium, it should be evident that with only a small extension of the preceding ideas, multicolor wavefront reconstruction should be possible.

The basic idea of color holography is to record three (or more) separate holograms on a single photographic plate (or other suitable recording medium), each with a different color, in such a way that subsequent illumination with a three-color beam yields three separate wavefronts, one in each of the primary colors, representing the portion of the object corresponding to that color.

Multicolor wavefront reconstruction was first proposed by Leith and Upatnieks in one of their original papers [1]. Their technique is the most straightforward of all: the hologram records three incoherently superposed diffraction patterns on a photographic emulsion. Each component hologram is a record of the object as it would appear when illuminated with a single color. If the object and reference beams contain the three primary colors, illumination of the composite hologram with a beam identical to the reference beam yields a reconstructed wavefront which closely approximates the wavefront that would result if the object were illuminated with white light. There are several other wavefronts which are produced in the process, however, and these overlap and generally degrade the color image. These spurious wavefronts, leading to "cross talk" images, result from light of wavelength λ_1 diffracting from the component hologram recorded with the wavelength λ_2. In this method, there will be six spurious, primary images. All of the methods of color holography to be discussed in this chapter represent various schemes to eliminate these crosstalk images. The technique described by Leith and Upatnieks consists of introducing the reference beam for each primary color at a different angle, thus spatially separating the crosstalk images from the true one.

Later, Mandel [2] noted that the crosstalk images are spatially removed from the desired image by a small amount due to the slightly different angles of diffraction for each color. Thus by viewing the image over a limited field, the reference beams for the three colors need not be introduced at different angles.

Pennington and Lin [3] utilized the thickness of the photographic emulsion to separate the crosstalk images from the desired ones. By using a thick emulsion, the hologram becomes essentially a three-dimensional diffraction grating, diffraction from which is governed by the Bragg relation

$$2d \sin \theta = \lambda \qquad (7.1)$$

where d is the grating spacing, θ the angle of incidence (and diffraction), and λ the wavelength. Figure 4.9 indicates that for small d (large values of θ_R in Fig. 4.9) and a thick emulsion, only a slight change in λ is required to extinguish the image. Thus for multicolor holograms recorded on a thick emulsion, the crosstalk images are largely suppressed since the illuminating wavelengths do not satisfy (7.1).

Later, Lin et al. [4] produced two-color holograms of the volume reflection type which yielded two-color images when illuminated with white light. As discussed in Chapter 4, when the object and reference beams enter the recording medium from opposite directions, standing waves are set up which result in a set of reflecting planes lying generally parallel to the emulsion surface. Since these reflecting planes are separated by $\lambda/2$, the hologram acts as a spectral filter, and the crosstalk images will again be suppressed. Because of the filtering action of the hologram, multicolor images may be obtained by illuminating with white light. The hologram will have high reflectance (diffraction) for only those wavelengths used to record the hologram.

Still later, Collier and Pennington introduced two more methods for making multicolor holograms without the requirement of a three-dimensional recording medium [5]. The first of these they called spatial multiplexing, in which the hologram is formed in such a way that the various component holograms in each color are not allowed to overlap on the emulsion. In reconstruction the hologram is illuminated so that each color illuminates only the portion of the hologram corresponding to that color. In this way, no crosstalk images are produced.

The second method requires coding of the reference beam so that each component reference wave varies in a unique manner over the hologram plate. In this case the amplitude and phase of the reference wave are made to vary across the hologram plate in a significantly different manner for each of the colors used to form the hologram. To reconstruct, it is necessary to relocate the hologram in exactly the position in which it was recorded and to illuminate it with a beam identical to the original coded reference beam. In this way the crosstalk images are sufficiently suppressed to render a good three-color image.

These six methods represent the typical schemes for recording multicolor holograms and the list is not meant to be exhaustive. Some of the proposed techniques work independently of the thickness of the recording medium, whereas the others demand that the thickness of the recording medium be large compared to the recorded fringe spacing. The aim of each technique is to suppress the unwanted crosstalk images. How this is achieved in each case is discussed in the following section.

7.1 ANALYSIS

Consider a multicolor wavefront that has been scattered by an object as given by

$$O(x, = \sum_{i=1}^{n} O_i(x) \tag{7.2}$$

and a multicolor reference wavefront

$$R(x) = \sum_{i=1}^{n} R_i(x). \tag{7.3}$$

Here $R_i(x)$ and $O_i(x)$ represent the complex amplitude at the hologram (x) plane corresponding to the wavelength λ_i.

Since there is no mutual coherence between the waves of different colors, the waves interfere at the hologram plane in such a way as to yield an irradiance

$$|H(x)|^2 = \sum_{i=1}^{n} [|O_i(x)|^2 + |R_i(x)|^2 + O_i^*(x)R_i(x) + O_i(x)R_i^*(x)]. \tag{7.4}$$

We suppose that the resulting amplitude transmittance of the hologram is simply equal to $|H(x)|^2$. Thus when the hologram is illuminated with a multicolor wavefront given by

$$C(x) = \sum_{j=1}^{n} C_j(x) \tag{7.5}$$

the resulting transmitted wave is given by

$$\psi(x) = \sum_{j=1}^{n} C_j(x) \sum_{i=1}^{n} (|O_i|^2 + |R_i|^2 + O_i^* R_i + O_i R_i^*). \qquad (7.6)$$

The first two terms in the parentheses on the right represent the n^2 zero orders and n^2 flare terms. The last two terms give rise to n^2 conjugate waves

$$\sum_{i,j}^{n} C_j O_i^* R_i \qquad (7.7)$$

and n^2 primary waves

$$\sum_{i,j}^{n} C_j O_i R_i^*. \qquad (7.8)$$

The terms in these sums for which $i \neq j$ yield the unwanted crosstalk images, whereas the terms for $i = j$ yield the true multicolor images.

To simplify the analysis, let us assume that the object, reference and illuminating beams originate from point sources. The multicolor object point is located at (x_o, y_o, z_o) and radiates at wavelengths λ_i. The reference point sources are located at $(x_R, y_R, z_R)_i$ and radiate at wavelengths λ_i. The subscript on these position coordinates indicates that each of the n wavelengths of the reference beam may originate from a different point. Similarly, the illuminating wave originates at point sources located at $(x_c, y_c, z_c)_j$ and radiate the wavelengths λ_j.

We may now extend (5.130) to include multiple wavelengths. By assuming no hologram scaling ($m = 1$), we obtain for the phases of the primary wavefronts (to first order)

$$\Phi_{P_{ij}}^{(1)} = \frac{2\pi}{\lambda_j} \cdot \frac{1}{2} \left[(x^2 + y^2) \left(\frac{1}{z_{cj}} + \frac{\mu_{ij}}{z_o} - \frac{\mu_{ij}}{z_{Rj}} \right) - 2x \left(\frac{x_{cj}}{z_{cj}} + \mu_{ij} \frac{x_o}{z_o} - \mu_{ij} \frac{x_{Rj}}{z_{Rj}} \right) \right.$$
$$\left. - 2y \left(\frac{y_{cj}}{z_{cj}} + \mu_{ij} \frac{y_o}{z_o} - \mu_{ij} \frac{y_{Rj}}{z_{Rj}} \right) \right]. \qquad (7.9)$$

Here $\mu_{ij} = \lambda_j/\lambda_i$ is the ratio of illuminating to reference wavelengths. To simplify the analysis still further, let us omit the y-dependence and just consider the terms in x. Each of the n^2 wavefronts of (7.9) yield a Gaussian image point located at

$$\mathbf{Z}_{P_{ij}} = \frac{z_{cj} z_j z_{Rj}}{z_o z_{Rj} + \mu_{ij} z_{cj} z_{Rj} - \mu_{ij} z_{cj} z_o} \qquad (7.10)$$

and

$$a_{P_{ij}} = \frac{x_{cj} z_o z_{Rj} + \mu_{ij} x_o z_{cj} z_{Rj} - \mu_{ij} x_{Rj} z_{cj} z_o}{z_o z_{Rj} + \mu_{ij} z_{cj} z_{Rj} - \mu_{ij} z_{cj} z_o}. \qquad (7.11)$$

Thus there are, in general, n^2 Gaussian image points and the problem is to separate the $n(n - 1)$ unwanted ones from the n desired ones and, moreover, have the n desired points coincident.

To do this, Leith and Upatnieks [1], using red, green, and blue wavelengths, separated the three reference sources so that

$$(x_R, z_R)_1 \neq (x_R, z_R)_2 \neq (x_R, z_R)_3 . \tag{7.12}$$

By illuminating the hologram with an identical wave,

$$(x_c, z_c)_j = (x_R, z_R)_j; \qquad j = 1, 2, 3, \tag{7.13}$$

they obtained, for example,

$$\mathbf{Z}_{P_{11}} = z_o = \mathbf{Z}_{P_{22}} = \mathbf{Z}_{P_{33}} \tag{7.14}$$

and

$$\mathbf{Z}_{P_{12}} = \frac{z_{c_2} z_o}{z_o(1 - \mu_{12}) + \mu_{12} z_{c_2}} \tag{7.15}$$

and also

$$a_{P_{11}} = x_o = a_{P_{22}} = a_{P_{33}}, \tag{7.16}$$

$$a_{P_{12}} = \frac{x_{c_2} z_o(1 - \mu_{12}) + \mu_{12} x_o z_{c_2}}{z_o(1 - \mu_{12}) + \mu_{12} z_{c_2}}. \tag{7.17}$$

From these examples we see that the mixed terms yield Gaussian image points that are spatially separated from the desired multicolor image point. Obviously, it is desirable to have $\mu_{ij} \, (i \neq j)$ large so that the crosstalk images are well separated from the true images.

When the thickness of the recording medium is comparable to or greater than the fringe spacing, the amount of diffracted image flux is strongly sensitive to the orientation of the hologram and to the wavelengths of the illuminating beam. With a thick hologram, then, it is much easier to separate the crosstalk images from the true image. Maximum flux is diffracted when the orientation of the three-dimensional fringe system satisfies (7.1). The grating spacing d is actually the separation between the planes through the recording medium. These planes act as if they were mirrors to the illuminating light when (7.1) is satisfied. The production of these planes was discussed in Chapter 4. If the two interfering waves contain several spectral components, a set of these reflecting surfaces (Bragg planes) are produced for each component. When this three-dimensional grating is illuminated with a plane wave containing the several spectral components, each component is diffracted according to (7.1), resulting in a plane wave of the same spectral composition. This is the reconstructed multicolor wavefront. However, there will also be several diffracted waves resulting from, for example, wavelength component λ_1 diffracting from the grating produced by component λ_2. This gives rise to the unwanted crosstalk images. For the three dimensional grating, however, these waves are much reduced in irradiance, as indicated in Fig. 4.9. Here we see that if the various

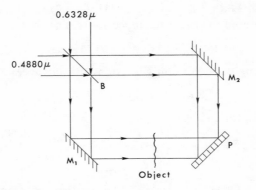

Fig. 7.1   Arrangement used by Pennington and Lin to record a two-color hologram [3].

spectral components are widely separated in wavelength, the recording medium relatively thick, and the grating spacing small, there will be very little flux diffracted into the unwanted directions. For a more complicated wavefront, the foregoing arguments still apply, since the complicated wavefront can be thought of as a superposition of many plane wavefronts. Thus the three-dimensional hologram is probably the most suitable for multicolor wavefront reconstruction.

The first such color hologram was reported by Pennington and Lin [3] of the Bell Telephone Laboratories. They reconstructed a two-color wavefront using the arrangement shown in Fig. 7.1. Light of wavelength 0.6328 μ from a He-Ne laser is mixed with blue light of wavelength 0.4880 μ from an Argon laser at the beamsplitter B. Part of this two-color wavefront proceeds along the upper path and constitutes the reference beam. The remaining part travels the lower path and passes through the object, which in their case was a color transparency. The two beams then interfere and expose the photographic plate P throughout the depth of the emulsion.

Subsequent illumination of the hologram with the blue wavefront alone reconstructed only the wavefronts which the transparency originally transmitted at this wavelength. A similar result was obtained with the red light, indicating that good rejection of the crosstalk images was achieved.

Three-color wavefronts, yielding better color fidelity than the two-color wavefronts, have also been recorded [6]. The basic arrangement used is shown in Fig. 7.2. Two of the spectral components (0.488 μ and 0.514 μ) are derived from an Argon laser. These are mixed with a third component at 0.633 μ from a He-Ne laser at the beamsplitter B. The object beam illuminates a diffusely reflecting object, yielding a complex wave of three colors which interfere with the corresponding colors in the reference beam

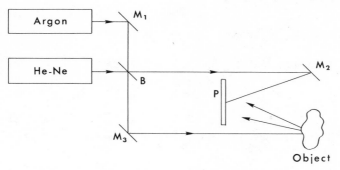

Fig. 7.2 An arrangement for recording a three-color hologram [6].

coming from mirror M_2. A thick emulsion photogaphic plate P is exposed to these interference patterns and becomes the hologram. If the angle between the reference and object beams is kept relatively large, the grating spacing will be small and crosstalk image rejection good. Better rejection of the crosstalk images can be achieved with a larger separation of the blue and green components.

In Fig. 4.13 we can see that the maximum crosstalk image suppression will be achieved when the angle between the multicolor reference and object beams is so large that they enter the recording medium from opposite sides ($\theta_R = 180°$). The Bragg planes then lie approximately parallel to the surface of the recording medium and the hologram is viewed in reflection. Holograms of this type were also discussed in Chapter 4 and the extension of the basic ideas to multicolor wavefront reconstruction is straightforward.

The major feature of this type of hologram is that a multicolor wavefront will be reconstructed with white-light illumination of the hologram. Since the Bragg planes lie approximately parallel to the emulsion, they are separated by approximately $\lambda/2n$, where n is the index of refraction of the emulsion and λ is the wavelength in air. These planes act as a spectral filter and have a high reflectivity only for those wavelengths which produced them. Hence if the object and reference beams are composed of several wavelength components, there will be several (one for each wavelength) sets of reflecting planes produced. Upon illumination of the resulting hologram with white light, the original spectral components are filtered out. A schematic of the formation and readout of this type of hologram is shown in Fig. 7.3. The multicolored reference beam is incident from one side of the recording medium P; the object beam enters the other side. Readout is accomplished by placing the eye at E and illuminating the hologram with either white light or a beam identical to the reference beam. A full-color image of the object can be observed. Illumination of the hologram with a

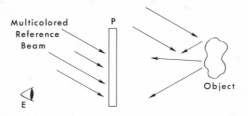

Fig. 7.3   Recording a multicolor hologram that can be viewed in reflection by illumination with white light.

beam the same as the reference beam is problematical if the hologram has been recorded on a photographic emulsion. Because of emulsion shrinkage, the Bragg planes will pull together so that the final spacing will no longer be $\lambda/2n$. Instead the spacing will be $\lambda'/2n$, with $\lambda' < \lambda$, so that little or no flux will be reflected at the original wavelengths. Because of this, white-light illumination is often more suitable, unless steps have been taken to prevent emulsion shrinkage or the hologram has been recorded on some other recording medium. In fact, for normal processing of normal photographic emulsions, the shrinkage is great enough ($\sim$ 15 or 20%) that the blue component will be shifted entirely out of the visible spectrum. Obviously, no color fidelity at all will be achieved unless steps are taken to reduce the shrinkage and the hologram is illuminated with white light. White-light illumination, however, does result in a loss of image resolution, since the hologram acts as a spectral filter with a bandwidth of typically 100A or so, which is too large for high-resolution imagery.

There are two additional methods for recording multicolor wavefronts without the requirement that the recording medium be thick [5]. Since the grating spacing to thickness ratio need not be small, small angles between reference and object beams may be used, and hence lower resolution recording media are suitable. Further, these methods make possible the recording of multicolor wavefronts with surface relief effects, such as produced with photographic emulsions and thermoplastics.

The first of these two methods has been called nonoverlapping spatial multiplexing. By this method the crosstalk images are eliminated by recording the hologram in such a way that the various colored interference patterns are not allowed to overlap on the recording medium. The resulting hologram is thus really a composite of many nonoverlapping component holograms, each recorded in only a single color. To reconstruct the object wavefront, the hologram is illuminated in such a way that each component hologram is illuminated with only that color light used to form it. In this way the crosstalk images are completely suppressed.

There are many possible experimental arrangements for producing such a hologram. One possible scheme for recording a three-color wavefront is indicated in Fig. 7.4. The incoming three-color beam is intercepted by a mirror M_1 and a colored, diffusely reflecting object. The reference beam coming from the mirror passes through a color filter mask before interfering with the object beam at the hologram plane H. The mask consists of narrow strips of red, green, and blue filters. Each filter allows only one color to pass through it, resulting in narrow strips of red, green, and blue light incident on the hologram plane. The light in each strip then interferes with the light of the same color coming from the object, yielding a nonoverlapping multiplicity of single-color holograms. Illumination of the hologram through this same filter mask reconstructs the object wavefront, with no crosstalk images. Of course, this scheme results in some noise being produced because, for example, in a red component hologram, green and blue light from the object will also be present, which merely decreases the average transmission of this component hologram. Since the same will be true for each component hologram, the composite hologram will have only a limited fringe modulation, resulting in a low diffraction efficiency, and hence there will be a lower signal-to-noise ratio in the image. There are other possible arrangements that can be used to avoid this drawback.

Holograms have been successfully recorded using the nonoverlapping spatial multiplexing technique on photographic emulsions having much lower resolution capabilities than the commonly used Kodak Spectroscopic Plate, Type 649F. Reference-to-object beam angles as low as 15° were used, and the resulting multicolor images were free of crosstalk [5].

The second method for recording multicolor wavefronts on thin recording media has been called reference beam coding [5]. With this method the amplitude and phase of the reference wave are made to vary across the hologram plane in a manner which is different for each wavelength. After the hologram has been exposed and processed, it must be replaced in precisely the same position it occupied during exposure and illuminated with

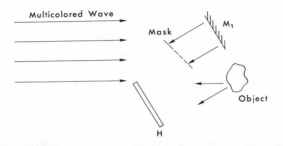

Fig. 7.4 Method for recording a three-color wavefront–nonoverlapping spatial multiplexing [5].

a beam which is identical in every respect to the original reference beam. If the illuminating beam is not identical to the reference beam, the object wave will not be reconstructed and the viewer will see only a relatively uniform noise distribution.

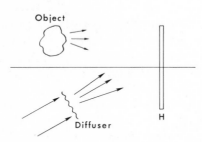

Object

Diffuser

H

Fig. 7.5 Method for recording a three-color wavefront-reference beam coding [5].

Figure 7.5 indicates a possible arrangement for recording this type of hologram. The multicolor reference wave is passed through a diffuser, such as a ground glass plate. As the wavefront progresses from the coding plate to the hologram plane, each of the color components is dispersed in a different manner. The complicated phase variations imposed on each wave as it passes through the code plate are further scrambled by the dispersion process in a manner which is strongly dependent on the wavelength. Hence each component reference wave is coded in a complicated manner. As an example of the wavelength sensitivity of the process, Collier and Pennington [5] have noted that when only light of wavelength 0.5145 μ was used, an excellent image was obtained. By changing the wavelength by only 0.0128 μ, however, and illuminating the coding plate in exactly the same way, no image at all was obtained.

Successful holograms have been recorded using this method, but there is still a noise problem. The noise results from what would normally produce the crosstalk images—the interaction of wavelengths λ_2 and λ_3 with the hologram formed with wavelength λ_1. The geometry of the code plate strongly affects the degree to which the noise interferes with the image. If the solid angle subtended by the coding plate at the hologram is small, the noise is fairly localized and therefore most annoying. The noise can be distributed more or less uniformly by increasing the solid angle subtended by the code plate.

The resolution obtainable with this method should be quite high, as discussed in Section 5.1. Since the reconstruction process represents an autocorrelation of the reference source distribution, the resolution can be quite high under the same conditions which give the least amount of crosstalk.

7.2 COLORIMETRY

The colors that can be reproduced holographically depend upon the choice of the three primary colors. These in turn depend upon the choice of lasers to be used. Not all lasers are suitable for color holography because of considerations of power at the desired wavelength and coherence length. For the sake of convenience, a power of 5 to 10 mw should be available at the desired wavelengths. This keeps the exposure times down to the order of a few minutes with the commonly used Kodak Spectroscopic Plate, Type 649F, even for relatively large objects.

The required coherence length is determined by the size of the object and the size of the hologram to be recorded. The most common choice of lasers for color holography today is the He-Ne laser with emission at 0.633 μ and the Argon laser with lines at 0.488 μ and 0.514 μ. All of these lines can be produced with ample power, but the Argon laser has a relatively short ($\sim$8 cm) coherence length. Various ways have been proposed for increasing the coherence length [7] and single mode oscillation at 0.514 μ has been attained [8].

The colors that can be reproduced with these three wavelengths can be determined by plotting them on the standard C.I.E. chromaticity diagram, shown in Fig. 7.6. This diagram gives the relative proportion x of one primary as the abscissa and that of another primary y as the ordinate. All

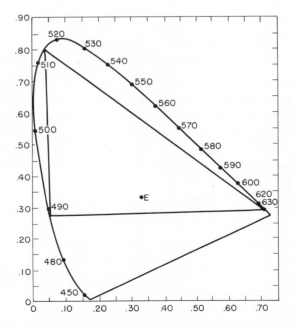

Fig. 7.6   The standard C.I.E. chromaticity diagram. The point E represents the chromaticity of a source radiating equal flux per unit bandwidth throughout the visible spectrum. The colors within the triangle can be reproduced with mixtures of the three wavelengths 0.488μ, 0.514μ, and 0.633μ.

visible colors are contained within the boundaries defined as the spectrum locus, consisting of the monochromatic colors from 0.700 μ down to 0.400 μ, and the line of purples, extending from 0.400 μ to 0.700 μ.

When any two colors are mixed, the new color lies somewhere along the line joining the two mixed colors in the diagram. Since the shape of the diagram is essentially triangular, it can be seen that three suitably chosen colors can be mixed in various proportions so as to yield a large range of colors but not all. For the wavelengths 0.488 μ, 0.514 μ, and 0.633 μ used for most color work, any color within the triangle indicated in Fig. 7.6 can be reproduced. It is seen that the deep blues and saturated (near the spectrum locus) yellows and greens will not be reproduced.

Table 7.1

Wavelength (Å)	Laser	Power (mw)	Coherence Length (cm)
4762	Krypton	25	12
4880	Argon	1000	8
5145	Argon	1000	8
5208	Krypton	50	12
5682	Krypton	50	12
6328	He-Ne	100	20
6471	Krypton	150	12

There are, however, several other laser lines that could prove to be more suitable for color holography. These are listed in Table 7.1, which is an incomplete listing of the laser wavelengths available for holography. The wavelengths indicated cover the visible spectrum quite well. The lasers producing these lines are all c.w. gas lasers. The coherence length is simply calculated as $c/\Delta\nu_D$, where c is the speed of light and $\Delta\nu_D$ is the Doppler width of the line.

REFERENCES

[1] E. N. Leith and J. Upatnieks, *J. Opt. Soc. Am.*, **54**, 1295 (1964).
[2] L. Mandel, *J. Opt. Soc. Am.*, **55**, 1697 (1965).
[3] K. S. Pennington and L. H. Lin, *Appl. Phys. Letters*, **7**, 56 (1965).
[4] L. H. Lin, K. S. Pennington, G. W. Stroke, and A. E. Labeyrie, *Bell System Tech. J.*, **45**, 659 (1966).
[5] R. J. Collier and K. S. Pennington, *Appl. Opt.*, **6**, 1091 (1967).
[6] A. A. Friesem and R. J. Fedorowicz, *Appl. Opt.*, **6**, 529 (1967).
[7] L. H. Lin and C. V. LoBianco, *Appl. Opt.*, **6**, 1255 (1957).
[8] J. M. Forsyth, *Appl. Phys. Letters*, **11**, 391 (1967).

8 Applications for Holography

8.0 INTRODUCTION

If holography is to become a viable, useful tool in modern industry, there must be some tasks for which it is the most suitable tool in a practical, not fundamental sense. The initial flurry of activity in the field after its introduction in 1948 and the vast amount of activity following the renaissance of holography in 1962 indicates that there are many people who feel that the holographic technique should prove quite useful. Indeed there have been myriad proposed uses for the technique, initially mainly concerned with microscopy but later expanded to include interferometry, information storage, character recognition, and many others. A few applications, proposed mainly by newswriters, must be classified as science fiction. Other proposed applications, such as information storage and character recognition, represent new ways of doing old jobs, and the task of proving that the new way is better or more practical than the old is not an easy one. Finally, there are several applications of the holographic technique that do represent true innovations. These are the applications that will eventually move into the forefront. They represent true innovations primarily because they utilize and exploit the fundamental difference between holography and photography—a hologram is a recording of a wavefront, whereas a photograph is a recording of the irradiance distribution in an image.

Further, there are different forms of holography which permit recording of previously unattainable information. These include x-ray and electron holography and, more recently, ultrasonic holography.

As long as some of the proposed applications prove feasible, people and industry will continue to work in the field, with the result that more and more applications should be forthcoming. We will discuss in this chapter several of the most interesting proposals. Some of these do not represent unique solutions to the problem at hand but nevertheless appear promising as better ways. Other proposed applications are unique in that there are no other means for performing the task. One of the most important in this latter category is holographic interferometry, with which we shall begin.

8.1   HOLOGRAPHIC INTERFEROMETRY

8.1.1   Introduction

Ever since the wave nature of light was generally accepted, interferometry has been the primary method for making measurements with great accuracy. The very small wavelength of light, of the order of 5×10^{-5} cm, and the fact that interferometric means are available for detecting changes of only a small fraction of this length, indicates the degree of accuracy which can be achieved. The very widespread applications of the method attest to its general usefulness. Interferometry is used for testing optical components, optical gauging of machine tools, studying air flow in wind tunnels, and standardizing the fundamental units of length. Therefore it is understandable that any fundamental improvement or innovation in present interferometric technique would find many applications over a wide field.

Holographic interferometry is such a fundamental innovation. Holography has widened the scope of interferometry to such a degree that it is hard to believe that holographic interferometry will not be used as a standard tool in engineering laboratories all over the world in a few years.

Conventional interferometry can be used to make measurements on highly polished surfaces of relatively simple shape. Holographic interferometry extends this range by allowing measurements to be made on three-dimensional surfaces of arbitrary shape and surface condition. A roughly machined machine part can now be measured to optical tolerance. Furthermore, with the holographic technique a complex object can be examined interferometrically from many different perspectives, because of the three-dimensional nature of the hologram. A single interferometric hologram is equivalent to many observations with a conventional interferometer. This property is especially useful for observations of such things as fluid flow in a wind tunnel [1]. A third departure of holographic interferometry from conventional interferometry is that an object can be interferometrically examined at two different times; one can detect with wavelength accuracy any changes undergone by an object over a period of time. The present object can be compared with itself as it was at an earlier time. This is a very great advantage in many fields. A large lens can be tested before and after mounting, for example. With the use of pulsed lasers, a machine part can be interferometrically compared with itself statically and dynamically.

Many more specific applications for this powerful new technique could be mentioned, but we will instead discuss in more detail the major proposed applications and methods of holographic interferometry. Specifically these are single- and double-exposure holographic interferometry, vibration analysis, contour generation and pulsed laser interferometry.

8.1.2 Single-Exposure Holographic Interferometry

To understand the basic process of holographic interferometry, first consider a hologram of an object wave $O(x)$ recorded as shown in Fig. 8.1. The object wave at the hologram plane is denoted by $O(x)$. This can be written in the usual form

$$O(x) = O_o(x)e^{i\varphi_o(x)} \tag{8.1}$$

where $O_o(x)$ is the (real) amplitude distribution and $\varphi_o(x)$ is the phase distribution in H. The reference wave at the hologram is written

$$R(x) = R_o e^{ikx\sin\alpha_R} \tag{8.2}$$

with R_o being the amplitude of the reference wave at the hologram and α_R the off-axis angle. We assume that the exposure and resulting amplitude transmittance is simply $|H(x)|^2$, where $H(x) = O(x) + R(x)$, so

$$|H(x)|^2 = O_o^2(x) + R_o^2 + R_o O_o(x)[e^{i(\varphi_o(x)-kx\sin\alpha_R)} + e^{-i(\varphi_o(x)-kx\sin\alpha_R)}] \tag{8.3}$$

Next assume that we replace the hologram in exactly the same position it occupied during exposure and illuminate it with the wave

$$C(x) = O'(x) + R(x). \tag{8.4}$$

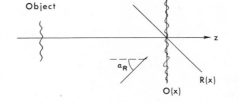

Fig. 8.1 Recording the object wave in the standard manner.

The wave $O'(x)$ means that we are illuminating the hologram with the object still in place, but we are allowing for a slight change in the object between recording the hologram and illuminating it. The wave $O'(x)$ is similar to $O(x)$ and we write it as

$$O'(x) = O_o(x)e^{i\varphi_o'(x)}, \tag{8.5}$$

that is, it has the same amplitude distribution but a slightly different phase distribution. The wave transmitted by the hologram becomes

$$\psi(x) = C(x) \cdot |H(x)|^2$$

$$= O_o^3 e^{i\varphi_o'} + O_o R_o^2 e^{i\varphi_o'} + R_o O_o^2 e^{i(\varphi_o - \varphi_o' - kx\sin\alpha_R)}$$

$$+ R_o O_o^2 e^{-i(\varphi_o - \varphi_o' - kx\sin\alpha_R)} + R_o O_o^2 e^{ikx\sin\alpha_R}$$

$$+ R_o^3 e^{ikx\sin\alpha_R} + R_o^2 O_o e^{i\varphi_o} + R_o^2 O_o e^{-i(\varphi_o - 2kx\sin\alpha_R)}. \tag{8.6}$$

This transmitted field consists of many waves traveling in several directions; the direction is determined by the linear variations of x in the exponents. The first, second, and seventh terms on the right-hand side of (8.6) represent component waves traveling in the approximate direction of the original object wave. These terms represent the new object wave $O_o(x)e^{i\varphi_o'(x)}$ which is transmitted directly through the hologram (first and second terms) and the reconstructed object wave (seventh term). Considering these terms alone, we have a transmitted component

$$\psi'(x) = O_o(x)\{[O_o^2(x) + R_o^2]e^{i\varphi_o'(x)} + R_o^2 e^{i\varphi_o(x)}\}. \tag{8.7}$$

The irradiance that the eye or camera sees is given by

$$|\psi'(x)|^2 = O_o^2(x)\{[O_o^2 + R_o^2]^2 + R_o^4 + 2R_o^2[O_o^2 + R_o^2]\cos(\varphi_o' - \varphi_o)\} \tag{8.8}$$

The irradiance distribution consists of a series of fringes; the location of a fringe is defined by the locus of all points for which

$$\varphi_o'(x) - \varphi_o(x) = \text{constant}. \tag{8.9}$$

A dark fringe is produced whenever

$$\varphi_o'(x_l - \varphi_o(x) = (2m + 1)\frac{\pi}{2}, \qquad |m| = 0, 1, 2, \ldots. \tag{8.10}$$

Figure 8.2 indicates the two interfering wavefronts. To show that $\psi'(x)$ is approximately the original object wave, we write $e^{i\varphi_o'(x)} = e^{i[\varphi_o(x) + \delta(x)]}$ so that (8.7) becomes

$$\psi'(x) = O_o(x)e^{i\varphi_o(x)}\{[O_o^2 + R_o^2]e^{i\delta(x)} + R_o^2\} \tag{8.11}$$

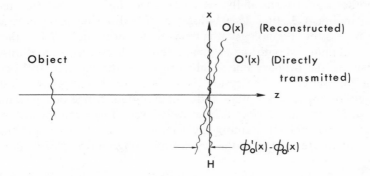

Fig. 8.2   The two interfering wavefronts in a holographic interferometer. The object, labeled Object′ in the figure, yields a wave $O'(x)$ at the hologram. This wave interferes with the reconstructed wave $O(x)$.

where $\delta(x)$ is the deviation of the changed object wavefront from the original. In most cases, $O_o^2(x)$ will be the rapid transmission variations resulting from the speckle pattern at the hologram from a diffuse object. Thus the wave $\psi'(x)$ just appears to be the original object wave modulated by transmission through this speckle pattern. Since $O_o(x)$ is already a rapidly varying function of x, the imposition of $O_o^2(x)$ on $O_o(x)$ does not yield a visible difference in the appearance of the image. If $\delta(x)$ is a smoothly and slowly varying function of x, the transmitted wave $\psi'(x)$ yields an image of the object crossed with fringes.

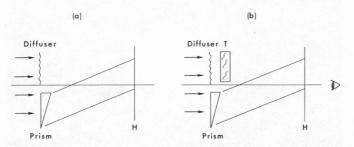

Fig. 8.3   A general purpose holographic interferometer. (a) A hologram of a diffuser is recorded. (b) The hologram is placed exactly in its original position so that both the reconstructed and actual object wave are superposed. When the test plate T is inserted, any optical inhomogeneities will distort the actual object wave so that interference fringes appear.

Another useful form of the single-exposure hologram interferometer is shown in Fig. 8.3. In 8.3*a* a hologram of a diffusing plate is recorded. After processing, the hologram is repositioned accurately. When the hologram is illuminated with the original reference beam, with the diffuser still in position, the primary image falls exactly in the position still occupied by the diffuser itself. The system is adjusted so that these two waves lie precisely on top of each other and a single bright fringe occupies the field of view. If a test flat T (Fig. 8.3*b*) now is inserted between the diffuser and hologram, but not intercepting the reference beam, fringes appear in the image which are related to the optical inhomogeneities, wedge and thick-

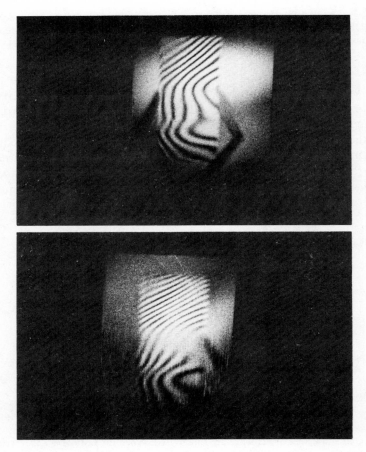

Fig. 8.4   Two views of an interferogram of a microscope cover slide made with the interferometer of Fig. 8.3.

ness variations present in the test plate. The operation of this sort of inter-ferometer is similar to the common Twyman-Green interferometer, except that the optical quality of T may be examined from many perspectives. Figure 8.4 shows a typical result which can be obtained using this type of interferometer. The test plate T is a microscope cover glass. The two photos show the resultant interference pattern for two perspectives.

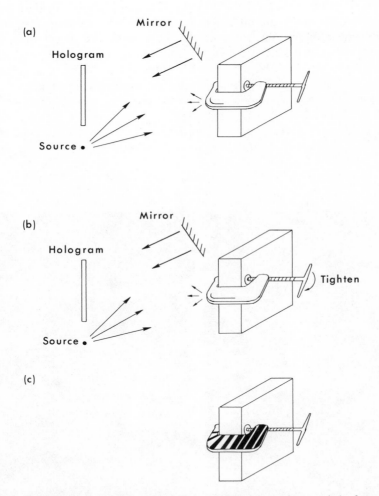

Fig. 8.5   Holographic strain interferometry. (a) A hologram is recorded of an object. (b) The hologram is replaced in its exact original position and the object is strained. (c) Light from the strained object passes directly through the hologram and interferes with the re-constructed wave of the unstrained object.

The single-exposure holographic interferometer may also be used to measure surface deformation in arbitrary objects at two different times [2]. A hologram of an object is made (Fig. 8.5a). The hologram is accurately replaced in its original position and the object is stressed. In 8.5b the hologram is illuminated with the reference beam and the light scattered from the stressed object. A view of the object through the hologram is shown in 8.5c. Interference fringes are present which are related to the degree of deformation of the object surface; quantitative calculations can be made of the exact amount of deformation present, although not simply. Figure 8.6 is an interferogram of this type, with the object under two amounts of strain. The stress was introduced in the object by tightening the c-clamp very slightly. This same technique can be used to measure surface deformations of all kinds, including thermal expansion and contractions, swelling caused by absorption, and any minute changes that might occur in any object.

The single-exposure hologram interferometer is particularly useful for observing phenomena within a transparent object, regardless of the optical

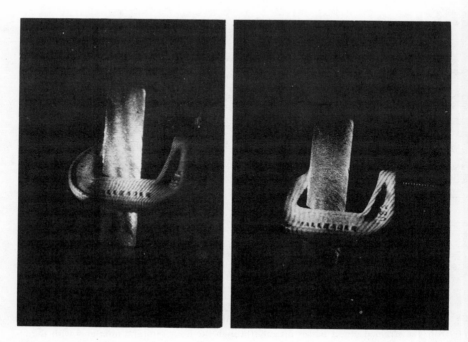

Fig. 8.6   Two different amounts of stress on a c-clamp, recorded as described by Fig. 8.5.

quality of the envelope material. A hologram is made with the object placed behind the diffusing plate of Fig. 8.3*a*. Then, if the object changes in any way, fringes will appear within the object. Suppose the object were a gas-filled lamp, for example. A hologram is made of the object with the lamp off. If the lamp is then turned on so that the gas heats up, fringes will be observed caused by the changes in optical path through the gas (see Fig. 8.9). The optical quality of the envelope is immaterial, since the object is being compared with itself. The optical path variations due to imperfections in the envelope are automatically cancelled. This would not be true for a conventional interferometer, where the object wave would be compared with a constant reference, such as a plane wave. In this case, fringes would result from the interference of the wave deformed by the envelope and the reference, resulting in an extremely complicated fringe pattern in which it would be difficult to distinguish between the permanent variations of the object and the variations induced by heating. For this reason holographic interferometry has been called differential interferometry—detecting only changes in an object.

8.1.3   Double-Exposure Holographic Interferometry

The double-exposure holographic interferogram is similar to the single-exposure case in most respects. Accurate registration of object and hologram is no longer required, however. Precision optical components are not required and it is still a differential interferometer; also a complete three-dimensional record of the interference phenomena is obtained, permitting post-exposure focusing and examination from various directions, just as in the single-exposure case.

The differences between single- and double-exposure holographic interferometry lie in the fact that with the double-exposure method the hologram retains as a permanent record the change in shape of the object between exposures. The two interfering waves may be reconstructed with a separate arrangement, without the need for an accurate registration of the plate in its original position. By using this technique, a permanent record can be made of differential changes in an object over a period of time.

The double-exposure interferogram requires making two holograms on a single recording medium. One of the two holograms yields a primary image which constitutes the comparison wave, just as in the single-exposure case. The test wave is not the object itself, however, but a reconstructed wave from the changed object. Interference phenomena, caused by changes in optical path through the object between exposures, are produced when the double hologram is illuminated.

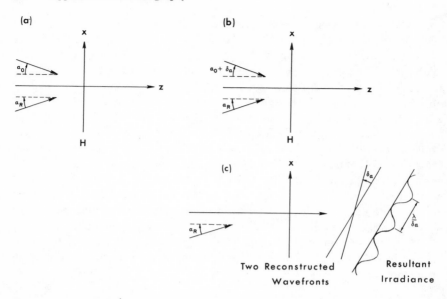

Fig. 8.7   A simple example of a double-exposure interferogram. (*a*) A hologram of a plane wavefront is recorded. (*b*) The hologram is doubly exposed by recording another hologram of a plane wavefront incident at a slightly different angle. (*c*) Both of these waves are reconstructed and they interfere so as to form straight fringes of spacing $\lambda/\delta\alpha$.

As a simple example of the double-exposure method, suppose we make a hologram of a plane wavefront, propagating at an angle α_o to the axis (Fig. 8.7*a*). The reference wave, also plane, is traveling at an angle α_R to the axis. The irradiance at the hologram plane H is

$$|H|^2 = |e^{iax} + e^{ibx}|^2 = e^{i(a-b)x} + e^{-i(a-b)x} + 2 \tag{8.12}$$

where $a = k \sin \alpha_o$ and $b = k \sin \alpha_R$; the amplitudes of the object and reference waves are taken as unity. Now suppose that a second exposure is made on the same recording medium, only now the object beam is incident at a slightly different angle, $\alpha_o + \delta\alpha$ (Fig. 8.7*b*). This hologram exposure is

$$|H'|^2 = |e^{i(a+\epsilon)x} + e^{ibx}|^2 = e^{i(a+\epsilon-b)x} + e^{-i(a+\epsilon-b)x} + 2 \tag{8.13}$$

where $\epsilon \cong k(\delta\alpha) \cos \alpha_o$ for $\delta\alpha$ small. If we now reconstruct by illuminating with the reference wave e^{ibx}, the transmitted wave becomes

$$\psi(x) = 4e^{ibx} + e^{iax} + e^{-i(a-2b)x} + e^{i(a+\epsilon)x} + e^{-i(a+\epsilon-2b)x}. \tag{8.14}$$

The second and fourth terms on the right-hand side represent the two reconstructed primary wavefronts (Fig. 8.7c):

$$\psi_P(x) = e^{iax} + e^{i(a+\epsilon)x}. \tag{8.15}$$

A detector placed behind the hologram will detect an irradiance proportional to

$$|\psi_P(x)|^2 = 2 + 2\cos(\epsilon x), \tag{8.16}$$

so that there are fringes in the transmitted wave representing the displacement of the object wave. The fringe spacing along the tilted wavefront is given by

$$\Delta(x\cos\alpha_o) = \frac{\lambda}{\delta\alpha}. \tag{8.17}$$

Another way of interpreting the fringes is as a spatial beat frequency. Each hologram consists of a single spatial frequency exposed onto the recording medium; the second spatial frequency is a little larger than the first. The composite hologram thus consists of a fringe pattern which is periodically washed out and reinforced. In the region of the washed-out fringe pattern, the wavefront does not reconstruct, giving rise to the modulated wavefront on the exit side of the hologram. One way to observe this spatial beat frequency would be to incoherently illuminate the hologram and then look at the hologram itself. Dark bands, representing the spatial beats, or moiré pattern, are then observable in the regions where the grating pattern was washed out by path differences of an odd number of half wavelengths.

The operation of the double-exposure hologram interferometer with diffuse objects is exactly analogous to the single-exposure case. Conceptually, one can extend the foregoing ideas for the plane wave case to the diffuse object wave case by considering the object wave as a set of plane waves propagating in various directions. The composite hologram will again contain the moiré beats between the two hologram exposures, resulting in the fringes on the reconstructed waves.

The double-exposure technique is well suited to interferometric recording of transient phenomena, such as shock waves and fluid flow, when a pulsed laser is used as the source. All of the principles discussed thus far apply equally well to time-dependent events, and the very short pulse of light from a ruby laser can record the interference phenomena at a single instant of time. The very wide range of applicability of the method has been well demonstrated by Brooks, Wuerker, Heflinger, and their co-workers at TRW, Inc. [3, 4, 5, 6]. Two excellent examples of their work are shown in Figs. 8.8 and 8.9.

Fig. 8.8   A double exposure holographic interferogram of the shock wave of a .22 caliber bullet traveling at 3,500 ft./sec. One exposure was made before the bullet was fired and the second just as the bullet and shock wave passed into view. The hologram was recorded using a Q-switched ruby laser and illuminated with the light from a He/Ne laser. A diffuser behind the object permitted it to be viewed from different angles, although the photo is taken at only one point of view. (After Brooks, Wuerker, Heflinger and Knox [6].) (Courtesy TRW, Inc.)

Although we have only discussed double exposures in this section, it is known that this technique can be extended to multiple exposures, with some interesting results [7]. Suppose, for example, that the object changes by some small increment between each exposure. The holographic image will contain fringes corresponding to the change of shape of the object, but the bright fringes will be much sharper than in the double exposure.

Fig. 8.9   The double exposure holographic interferogram as a differential interferometer. One exposure was made with the filament cold and the second with the filament heated. The difference between the two objects is only the density change of the filling gas. The interference pattern describing this change is accurately displayed interferometrically in spite of the poor optical quality of the envelope material. (After Brooks, Wuerker, Heflinger and Knox [6].) (Courtesy TRW, Inc.)

Multiple-exposure holographic interferometry, as with ordinary multiple-beam interferometry, yields sharp fringes, so that fringe displacement may be measured with great accuracy.

8.1.4   Time-Average Holographic Interferometry

The idea of multiple-exposure interferometry may be extended to the limiting case of a continuum of exposures, resulting in what might be called "time-average holographic interferometry." This technique lends itself very nicely to the problem of vibration analysis and may very well be the best method yet devised for such analysis. The problem of recording a hologram in the presence of subject motion has been discussed in detail in Section 6.2, so only a qualitative description of the process will be given here. The earliest report of this form of holographic interferometry was by Powell and Stetson [8]. The basic idea of the method is that since holography itself is an interferometric process, any instabilities of the interferometer cause fringe motion. Thus the hologram of a vibrating object is a record of the time-averaged irradiance distribution at the hologram plane. Since the amount of light flux diffracted from any region of the hologram depends on the fringe contrast, any object motion that causes the fringes to move during the exposure, causing a loss in contrast, will result in less diffracted flux from that region of the hologram. The strength of the reconstructed wave is therefore a function of the fringe motion during the exposure. The process can be thought of as a recording of very many incremental holograms, one for each incremental position of the object. In the reconstruction, each incremental hologram yields an image wave, each slightly displaced, producing the interference effects.

If the object is vibrating in a normal mode, there will be standing waves of vibration on the surface, so that at the nodes the object motion will be very small or nonexistent. At the antinodes the vibration amplitude will be large. A hologram of such an object will then produce a bright image of the regions of the object for which little or no motion occurred during the exposure, whereas it will not produce images of the antinodal points at all. Figure 8.10 shows some of the exceptional photos taken by Powell and Stetson using this technique. The vibrating surface is the end of a cylindrical can. The nodes are the bright areas of the photo and by counting the contours lying between the nodes and a known stationary object point,

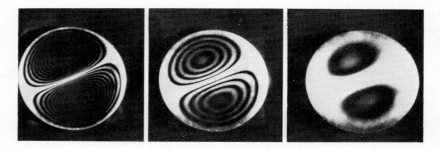

Fig. 8.10   Some of the time-average holographic interferograms produced by Powell and Stetson [8]. The vibrating surface is the end of a 35 mm film can.

the amplitude of vibration for each object point can be found, since each bright area represents an amplitude of vibration of approximately a multiple of a half wavelength, depending on the recording geometry.

This holographic method of vibration analysis has all of the advantages of holographic interferometry. The method can be used regardless of the shape or complexity of the object; the vibration modes can be examined in three dimensions, or at least from a variety of perspectives; and the method works regardless of whether the surface is optically smooth or diffusely reflecting. The technique should find widespread application.

8.1.5   Contour Generation

The idea of contour generation was first proposed by Haines and Hildebrand [9] and should find some interesting and useful applications. The basic idea behind the process can be easily explained. Suppose, as shown in Fig. 8.11a, a hologram is made of a single point object a distance z_o from the hologram plane H. The reference wave is a plane wave, making some angle with the hologram plane. Let us further assume that the light source used for this recording contains two wavelengths, λ_1 and λ_2. Since the light at these two wavelengths has no mutual coherence, the recorded hologram is really two incoherently superposed holograms of the point object. We now illuminate the hologram with light of wavelength λ_1 only. The illumi-

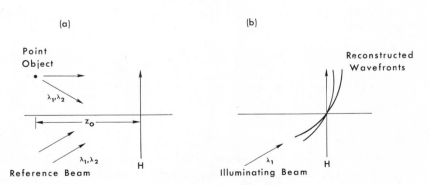

Fig. 8.11   Holographic contour generation. (a) Recording an object point with two wavelengths. (b) Reconstructing two wavefronts with different magnifications.

nating wave is also plane and is incident on the hologram at the same angle as the reference wave, so that a spherical wave is reconstructed which appears to have originated at the point at z_o. This reconstruction is produced by the component hologram recorded with the light of wavelength λ_1. The other component hologram, however, recorded with the light of wavelength λ_2, reconstructs another spherical wavefront, with a different radius (Fig. 8.11b), because of the magnification resulting from recording and reconstructing with different wavelengths. From (5.132) we see that the expression for the radius of a reconstructed spherical wavefront is

$$\mathbf{Z}_p = \frac{z_c z_o z_R}{z_o z_R + \mu z_c z_R - \mu z_c z_o} \tag{8.18}$$

where the hologram scaling factor m is unity and the wavelength ratio μ is the ratio of the wavelengths used for recording and illumination. In this case, $z_c = z_R = \infty$, so (8.18) becomes

$$\mathbf{Z}_p = \frac{z_o}{\mu}. \tag{8.19}$$

For the hologram recorded at λ_1, $\mu = 1$, and for the hologram recorded at λ_2, $\mu = \lambda_1/\lambda_2$. Hence the two spherical waves produced have radii z_o and z_o/μ, so that a viewer looking through the hologram sees two virtual images, one at a distance z_o from the hologram and the other at a distance z_o/μ. The optical path difference between these two wavefronts is

$$OPD = \Delta z_o = \frac{z_o}{\mu} - z_o = \frac{z_o}{\mu}(1 - \mu) = z_o \frac{\Delta \lambda}{\lambda_1} \tag{8.20}$$

where $\Delta\lambda = \lambda_1 - \lambda_2$. For most cases, $\Delta\lambda$ is chosen small enough so that the actual image observed appears to be a single point. Whether or not the point is visible at all, however, depends on how far the point is from the hologram. For values of z_o such that

$$\Delta z_o = m\lambda_1, \qquad |m| = 0, 1, 2, \ldots \tag{8.21}$$

a bright point will be observed. However, for values of z_o such that

$$\Delta z_o = (m + \tfrac{1}{2})\lambda_1 \tag{8.22}$$

the point will not be observable, because of destructive interference between the two waves. In the continuous case we see that for a continuous array of points in an object, only those portions of the object at a distance z_o from the hologram will appear in the image, where z_o is given by the condition

$$z_o \frac{\Delta\lambda}{\lambda_1} = m\lambda_1, \qquad |m| = 0, 1, 2, \ldots. \tag{8.23}$$

Therefore the image of an object recorded with two discreet wavelengths separated by $\Delta\lambda$ will appear to have fringes on the surface; each fringe is the locus of all points at a constant distance from the hologram. The fringes are contours of constant depth of the object, with the change in depth between fringes given by

$$\delta z_o = \frac{\lambda_1{}^2}{\Delta\lambda} \tag{8.24}$$

which follows from (8.23). Clearly a more rigorous analysis would include consideration of the viewing angle, since the optical path difference between the two waves will be a function of position in the hologram plane.

The occurrence of $\lambda^2/\Delta\lambda$ in (8.24), which is just the coherence length of the light used to make the hologram, indicates an interesting use of this technique. Suppose that the object is very long and positioned as in Fig. 8.12. If a hologram is made of this object, only those points along the object satisfying (8.23) will yield bright images—for other values of z_o, the corresponding image points will be less intense—the irradiance distribution in the image will be directly related to the coherence properties of the source. If the source is a single line of width $\Delta\lambda$, the image brightness will decrease from a maximum to zero over a distance $\delta z_o = \lambda^2/\Delta\lambda$. If the source contains two discreet lines separated by $\Delta\lambda$, the image will be fringed as previously discussed, since the fringe visibility function, as given just following (6.27), is periodic with period $\lambda^2/\Delta\lambda$. For more complicated distributions the irradiance distribution in the image is more complicated but is still proportional to the visibility function. Therefore we see that this technique may also be used as an interferometric spectrometer.

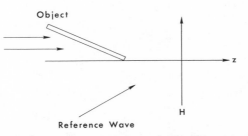

Fig. 8.12 A possible holographic arrangement for mapping the fringe visibility function of a light source.

8.2 CORRECTING ABERRATED WAVEFRONTS WITH HOLOGRAMS [10]

The use of extra optical devices in imaging systems to improve their performance is not uncommon; the most common image-improving device is the stop. Use has been made of Fresnel zone plates, phase plates, and other devices for the purpose of improving optical images. Holograms, too, may be employed for this purpose, but their use has several drawbacks which severely limit the usefulness of this technique. Nevertheless, some special cases may arise wherein the hologram technique could prove quite useful. Also, knowledge of the process of aberration correction with holograms is useful for anyone attempting to make holograms with good image resolution where lenses are used in making the hologram and forming the image, since it is possible at least partially to compensate for the lens aberrations with proper technique. The basic system used for making the hologram is shown in Fig. 8.13a. Suppose the lens L produces a wavefront W with aberrations. The reference wave R is plane. The wave W in the hologram plane is described by

$$O(x) = e^{i\varphi_0(x)} \tag{8.25}$$

and the reference wave by

$$R(x) = e^{ikx \sin \alpha_R} \tag{8.26}$$

as usual, where we are assuming unit amplitude. After processing, the amplitude transmission of the hologram is assumed to be simply

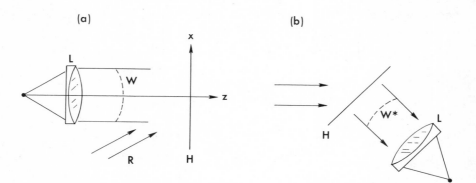

Fig. 8.13   Holographic correction of lens aberrations. (*a*) Recording the aberration. (*b*) The hologram produces a wave that is phase conjugate to the aberrated wave. When the lens L images this wave, the errors will cancel.

$$|H|^2 = |R + O|^2 = 2 + e^{i[\varphi_o(x) - kx\sin\alpha_R]} + e^{-i[\varphi_o(x) - kx\sin\alpha_R]}. \quad (8.27)$$

Since the lens L has been used to collimate the light, the function $\varphi_o(x)$ is just the phase difference between W and an ideal plane wavefront. The last term on the right of (8.27), however, contains a phase term which is conjugate to W—hence the hologram can be used as a corrector plate.

To do this, the hologram is illuminated from behind so that a wavefront W^* is produced (Fig. 8.13b). That is, the hologram is illuminated in such a way that all of the rays of 8.13a are exactly reversed. The lens L should then produce a diffraction-limited point image in its focal plane, since the incident wavefront W^* is expressed as $e^{-i\varphi_o(x)}$, which will just be cancelled when multiplied by the lens phase error, $e^{i\varphi_o(x)}$.

Figure 8.14 shows the results achieved by Upatnieks et al. [10] with this method. These are the point images produced by a poor lens (a) without the hologram corrector plate, (b) with the hologram corrector plate, and (c) a 25× enlargement of Fig. 8.14b. Note that this last is almost an Airy disk.

Although the foregoing analysis considered only a collimated input to the lens, there is also improvement over a finite field, as evidenced in the photographs of Fig. 8.15. In Fig. 8.15a the lens is used without the hologram corrector plate; in Fig. 8.15b the corrector plate is used. The improvement of the image quality over the field is obvious.

The drawbacks to the method include the fact that so little light is diffracted by the hologram into the desired order—only about 2 to 3% is typical. It should be possible to make phase holograms with more diffraction efficiency, however, which would overcome this problem. Another disadvantage of the method is that monochromatic light must be used because of the dispersion of the hologram, which limits the general usefulness of the method still more. Methods are available for achromatization, such as use of prisms or gratings, but the system rapidly becomes quite complex. Finally, the main drawback is that the hologram corrector plate introduces new aberrations of its own. As shown in Chapter 5, whenever the illuminating wave is not identical to the reference wave, aberrations are introduced. If the corrector plate has been made for use with a certain lens, and the lens subsequently used to image a scene, not all points of the scene will illuminate the hologram from the same direction. Thus aberrations will be added to the waves from these scene points. The analysis of [10] shows how these additional aberrations may be minimized but never eliminated. The authors show that, under optimum conditions, there will still be some residual astigmatism and distortion introduced by the hologram corrector plate. Hence the scheme of correcting lenses with holograms only corrects some aberrations at the expense of increasing others.

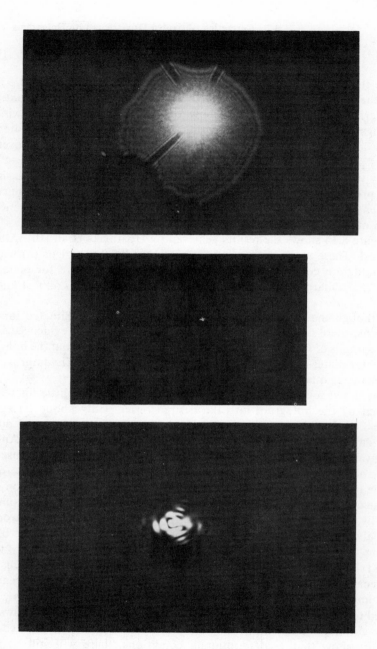

Fig. 8.14 Actual examples of holographic correction. (*a*) The point spread function of a lens with a large amount of spherical aberration. (*b*) The spread function of the lens when used with the holographic corrector plate. (*c*) A 25X enlargement of (*b*) [10].

Fig. 8.15 Holographic correction over an extended field. (*a*) Image using lens without corrector plate, (*b*) Image using lens with corrector plate [10].

8.3 MATCHED FILTERING AND CHARACTER RECOGNITION

The use of holograms as optical spatial filters was introduced by Vander Lugt [11] almost simultaneously with the introduction of holograms made with off-axis reference beams. Basically, the optical matched filter is a photographic recording of the Fourier transform of the amplitude distribution of an object. Since it is the amplitude distribution with which we are concerned, it is necessary to record the phase as well as the irradiance in the Fourier transform; this is why holograms are so well suited to this problem.

The important problem of character recognition is one for which holograms might be used advantageously, since optical character recognition is one of the prime uses for the optical matched filter. How the optical matched filter is used for character recognition can be described simply as follows: A number of masks are stored on a master hologram. Each mask corresponds to one of the characters to be read. An optical filtering process performs a cross-correlation between an unknown character and the master hologram. If the mask corresponding to the unknown character in the input plane is present, a bright spot appears in the output plane in a position which corresponds to that character.

The master hologram consists of a large number of Fourier transform holograms of a set of characters to be stored. It is produced as shown in Fig. 8.16. The field distribution in the x-y plane is the Fourier transform of the set of characters (Appendix A). The reference wave is plane and is denoted by R in Fig. 8.16.

The completed hologram is replaced as indicated in Fig. 8.17. When an unknown character is placed in the window in the x-y plane, the lens L

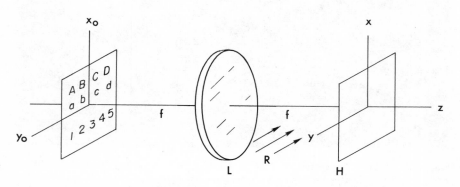

Fig. 8.16 Producing the master hologram for a holographic character recognition system.

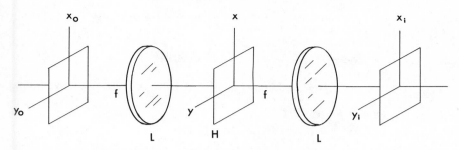

Fig. 8.17 Identifying an unknown character in the input plane with a holographic character recognition system.

produces its Fourier transform at the hologram. Here the Fourier transform of the unknown character is multiplied by all of the recorded Fourier transforms (by the process of transmission through the hologram). A second lens L produces a Fourier transform of this product in the x_i-y_i plane, and therefore the cross-correlations of the unknown character and all of the stored characters are displayed in this plane. Since the cross-correlation of two functions is the integral of their product as a function of relative displacement, a bright spot will appear in the x_i-y_i plane when the unknown character is correlated with a stored character of similar shape. This bright central spot corresponds to the integral over the product at zero displacement. This central bright spot will be surrounded by flare light, corresponding to the cross-correlations of the character to be read with all of the other stored masks. An array of detectors in the x_i-y_i plane would then determine the position of the brightest spot, which in turn identifies the unknown character.

The process may be described mathematically as follows: Suppose we make the mask of a single character $C(x_o)$. The total amplitude in the hologram plane (the x-y plane of Fig. 8.16) is then (in one dimension),

$$H(x) = R(x) + O(x) = R_o e^{ikx \sin \alpha_R} + \left(\frac{-i}{\lambda f}\right)^{1/2} \int_{-\infty}^{\infty} C(x_o) e^{(ik/f)xx_o} dx_o, \quad (8.28)$$

where R_o is the amplitude of the plane reference wave, which makes an angle α_R with the hologram normal. The recorded irradiance in this plane is

$$|H(x)|^2 = R_o^2 + |O(x)|^2 + R_o O(x) e^{-ikx \sin \alpha_R} + R_o O^*(x) e^{ikx \sin \alpha_R}.$$
$$(8.29)$$

It is the last term on the right which is of interest to us here, so we retain only that term. This term would normally lead to the conjugate image. Now let us suppose that the amplitude transmission of the processed hologram is simply equal to $|H(x)|^2$. The hologram is replaced in the x-y plane

(Fig. 8.17). If the character $C(x_o)$ now is placed in the aperture of the system $(x_o\text{-}y_o$ plane), a field

$$O(x) = \left(\frac{-i}{\lambda f}\right)^{1/2} \int_{-\infty}^{\infty} C(x_o')e^{i(k/f)xx_o'}\,dx_o' \qquad (8.30)$$

is incident on the hologram. The transmitted wave is then given by

$$\psi(x) = O(x)O^*(x)R_oe^{ikx\sin\alpha_R}$$

$$= \frac{R_o}{\lambda f}e^{ikx\sin\alpha_R}\int_{-\infty}^{\infty} C(x_o')e^{i(k/f)xx_o'}\,dx_o'\int_{-\infty}^{\infty} C(x_o)e^{-i(k/f)xx_o}\,dx_o. \qquad (8.31)$$

Combining these integrals yields

$$\psi(x) = \frac{R_o}{\lambda f}e^{ikx\sin\alpha_R}\iint_{-\infty}^{\infty} C(x_o)C(x_o')\exp\left[i\frac{k}{f}x(x_o' - x_o)\right]dx_o'\,dx_o. \qquad (8.32)$$

The second lens L of Fig. 8.17 now forms the image in the x_i plane:

$$G(x_i) = \left(\frac{-i}{\lambda f}\right)^{1/2}\int_{-\infty}^{\infty}\psi(x)e^{i(k/f)xx_i}\,dx. \qquad (8.33)$$

Substituting (8.32) into (8.33) gives

$$G(x_i) = \frac{R_o}{\lambda f}\left(\frac{-i}{\lambda f}\right)^{1/2}$$

$$\times \int_{-\infty}^{\infty}\left[e^{ikx\sin\alpha_R}\iint_{-\infty}^{\infty} C(x_o)C(x_o')e^{i(k/f)x(x_o'-x_o)}dx_o'\,dx_o\right]e^{i(k/f)xx_i}\,dx. \qquad (8.34)$$

By interchanging the order of integration we get

$$G(x_i) = \frac{R_o}{\lambda f}\left(\frac{-i}{\lambda f}\right)^{1/2}\iiint_{-\infty}^{\infty} C(x_o)C(x_o')$$

$$\times \exp\left[i\frac{k}{f}x(x_o' - x_o + x_i + f\sin\alpha_R)\right]dx_o\,dx_o'\,dx. \qquad (8.35)$$

By writing

$$\delta(a) = \frac{1}{2\pi}\int_{-\infty}^{\infty} e^{-iax}\,dx, \qquad (8.36)$$

and integrating over x, we obtain

$$G(x_i) = \frac{R_o}{\lambda f}\left(\frac{-i}{\lambda f}\right)^{1/2}2\pi\iint_{-\infty}^{\infty} C(x_o)C(x_o')$$

$$\times \delta\left[\frac{k}{f}(x_o' - x_o + x_i + f\sin\alpha_R)\right]dx_o\,dx_o'$$

$$= R_o\left(\frac{-i}{\lambda f}\right)^{1/2}\int_{-\infty}^{\infty} C(x_o)C(x_o - x_i - f\sin\alpha_R)\,dx_o. \qquad (8.37)$$

This is just the desired correlation function. It is represented by a bright spot in the x_i plane at the position $x_i = -f \sin \alpha_R$. To show that this position is a function of the position of the character during recording of the master hologram, we assume that the character was originally centered at a point a so that it is represented by $C(x_o - a)$ and

$$O(x) = \left(\frac{-i}{\lambda f}\right)^{1/2} \int_{-\infty}^{\infty} C(x_o - a) e^{i(k/f)xx_o} \, dx_o. \tag{8.38}$$

The term of interest in the hologram exposure thus becomes

$$R(x)O^*(x) = R_o e^{ikx \sin \alpha_R}$$
$$\times \left(\frac{-i}{\lambda f}\right)^{1/2} \int_{-\infty}^{\infty} C(x_o - a) \exp\left(-i\frac{k}{f}xx_o\right) dx_o. \tag{8.39}$$

When this same character is placed in the center of the aperture, the illumination of the hologram is

$$O(x) = \left(\frac{-i}{\lambda f}\right)^{1/2} \int_{-\infty}^{\infty} C(x_o') e^{i(k/f)xx_o'} \, dx_o' \tag{8.40}$$

and the transmitted wave becomes

$$\psi(x) = \frac{R_o}{\lambda f} e^{ikx \sin \alpha_R} \iint_{-\infty}^{\infty} C(x_o - a) e^{-i(k/f)xx_o} C(x_o') e^{-i(k/f)xx_o'} \, dx_o \, dx_o'. \tag{8.41}$$

Therefore the image is given by

$$G(x_i) = \left(\frac{-i}{\lambda f}\right)^{1/2} \int_{-\infty}^{\infty} \psi(x) e^{i(k/f)xx_i} \, dx$$

$$= \frac{R_o}{\lambda f} \left(\frac{-i}{\lambda f}\right)^{1/2} \iiint_{-\infty}^{\infty} C(x_o - a) C(x_o')$$
$$\times \exp\left[-i\frac{k}{f} x(x_o - x_o' - x_i - f \sin \alpha_R)\right] dx_o \, dx_o' \, dx$$

$$= \frac{R_o}{\lambda f} \left(\frac{-i}{\lambda f}\right)^{1/2} \cdot 2\pi \iint_{-\infty}^{\infty} C(x_o - a) C(x_o')$$
$$\times \delta\left[\frac{k}{f}(x_o - x_o' - x_i - f \sin \alpha_R)\right] dx_o \, dx_o'$$

$$= R_o \left(\frac{-i}{\lambda f}\right)^{1/2} \int_{-\infty}^{\infty} C(x_o - a) C(x_o - x_i - f \sin \alpha_R) \, dx_o$$

$$= R_o \left(\frac{-i}{\lambda f}\right)^{1/2} \int_{-\infty}^{\infty} C(y) C(y + a - x_i - f \sin \alpha_R) \, dy, \tag{8.42}$$

which has its maximum value at the point

$$x_i = a - f \sin \alpha_R. \tag{8.43}$$

Hence the bright spot representing the character originally described by $C(x_o)$ was located at $x_i = -f \sin \alpha_R$, whereas the spot representing the character $C(x_o - a)$ is located at $x_i = a - f \cdot \sin \alpha_R$. Thus the presence of a bright point in the output plane indicates the presence of a character in the aperture, and its position corresponds to the original position of the character when the master hologram was made, which identifies the character.

Burckhardt [12] has considered the problem of how many masks of individual characters can be stored on a single hologram. For a completely noiseless recording medium, he finds for the maximum number N of masks which can be stored

$$N = \frac{\pi \nu_c^2 D^2}{32 a^2 \nu^2}, \tag{8.44}$$

where ν_c = highest recordable spatial frequency
D = diameter of lens L in Fig. 8.16
a = dimension of input character
ν = maximum spatial frequency in the spectrum of the input character.

These N characters are stored on the master hologram with P separate exposures, where

$$P = \frac{N}{M_{max}} \tag{8.45}$$

and M_{max} is the greatest number of characters which may be stored in a single exposure:

$$M_{max} = \frac{\pi D^4}{256 f^2 \lambda^2 \nu^2 a^2}, \tag{8.46}$$

where f is the focal length of the lens L in Fig. 8.16, and λ is the wavelength of the recording light. M_{max} is a finite number because of the finite resolu- of the lens L. Assuming the following parameters

$$f = 5 \times 10^2 \text{ mm} \qquad D = 10^2 \text{ mm}$$

$$\lambda = 6.33 \times 10^{-4} \text{ mm} \qquad a = 3 \text{ mm}$$

$$\nu_c = 1.6 \times 10^3 \text{ }l/\text{mm} \qquad \nu = 10 \text{ }l/\text{mm},$$

we find that $N \approx 2.5 \times 10^6$ and $M_{max} \approx 10^4$.

For the case of a nonperfect recording medium, the maximum number of characters which may be stored is somewhat less for a given signal-to-noise ratio. The general equation in this case is given by

$$N = \frac{(T_o - 1)^2 F_1}{256(S/N)^2 R \cdot P} \cdot \frac{D^2}{a^2 v^2},\tag{8.47}$$

where T_o is the average amplitude transmission of the master hologram, F_1 is a constant (~ 30) which takes into account the fact that the area occupied by the correlation function (8.37) is not uniformly filled with light, S/N is the desired signal-to-noise power ratio, and R is the fraction of the incident power scattered per (unit bandwidth)^2. Taking $T_o = .5$, $F_1 = 30$, $S/N = 100$, $R = 6 \times 10^{-9}$, and $P = 1$, we find

$$N \approx 5 \times 10^3\tag{8.48}$$

This is, of course, smaller than that formed for the noiseless system, but is still sufficiently large to store all of the alpha-numeric characters, each with some 140 variants.

8.4 HOLOGRAPHIC MICROSCOPY

As mentioned in Chapter 1, holography was originally invented as an improvement of microscopy—specifically, electron microscopy. In the early years workers extended Gabor's ideas to include x-ray microscopy. Although neither of these could be made to come up to expectations, there is still a great deal of promise for holographic microscopy.

There are three main features that make the wavefront reconstruction method potentially more suitable for microscopy than conventional imaging techniques. The first of these is that, theoretically at least, many of the wave aberrations present in the recording process can be corrected by means of a suitable recording and readout arrangement. Thus the final image wave will be nearly free of aberrations, leading to almost a diffraction-limited image. The second feature is that in holography, the field of view is a function of the recording medium resolution and size. We can therefore expect, again theoretically, good imagery over much larger fields than are attainable with conventional microscopy. The third feature is, of course, that magnification is possible by employing a change in wavelength between recording and reconstructing the wavefront. Thus, as was originally proposed, we can record the wavefront in a very short wavelength and subsequently reconstruct it with a much longer one.

The main problem with holographic microscopy is that magnification is achieved by means of a wavelength and/or radius of curvature change between recording and reconstructing. These changes are always accompanied by the wavefront aberrations discussed in Section 5.4. The elimination and reduction of these aberrations was discussed there, and the holographic microscope has to be designed so as to eliminate most aberrations.

Generally speaking, holographic image resolution is limited by the diffraction limit of the hologram aperture, assuming, of course, that such things as bandwidth, source size, and aberrations are all made negligible, as discussed in Chapter 5. Therefore resolution in microscopy depends mainly on the wavelength used to make the hologram. This is why electron waves and x-rays appear so interesting.

The image magnification, on the other hand, depends on such things as the radii of curvature of the wavefronts used to record and illuminate the hologram, the ratio of the wavelengths used in recording and reconstructing, and the linear magnification of the hologram itself. The general formula for magnification was derived in Chapter 5:

$$M = \frac{m}{1 \pm m^2 z_0/\mu z_c - z_0/z_R},\qquad(8.49)$$

where the upper sign is for the primary image (magnification as viewed from the hologram if this image is virtual) and the lower sign for the conjugate image. The linear magnification of the hologram is denoted by m and the wavelength ratio by μ. As pointed out in Chapter 5, any magnification due to an increase of the illuminating wavelength must be accompanied by a corresponding scaling up of the hologram, otherwise image resolution will be lost by an increase in the aberrations.

A schematic of a possible holographic microscope is shown in Fig. 8.18a. The specimen O is placed a distance z_0 from the hologram H. For reconstruction (8.18b) the wavelength may be different from that used in recording, and the hologram (denoted by H') scaled accordingly. The

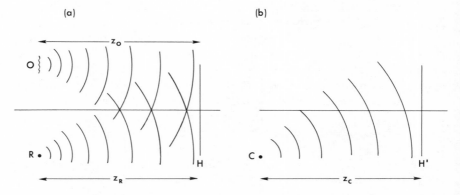

(a)   (b)

Fig. 8.18 A holographic microscope. (a) A hologram of the object O is recorded with a spherical reference wave derived from a point source a distance z_R from the hologram plane. (b) The hologram, which may have been scaled, is denoted by H' and is illuminated with a spherical wave derived from a point source a distance z_c from the hologram.

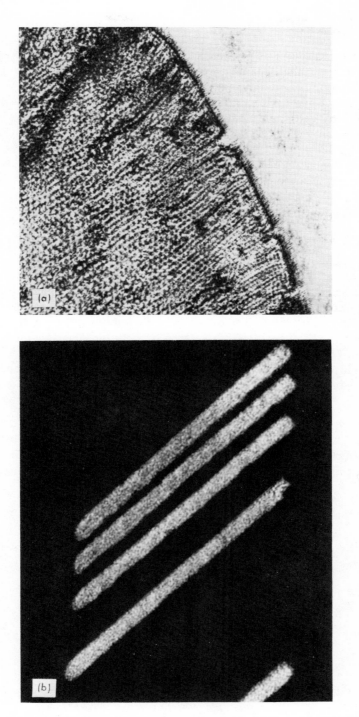

Fig. 8.19 Examples of holographic microscopy, after Leith and Upatnieks [13]. (*a*) Fly's wing, $M = 60$, $\lambda = 0.633\mu$, $m = \mu = 1$. (*b*) Bar pattern, $M = 120$, $\lambda = 0.633\mu$, $m = \mu = 1$. The spacing between the closest bars is $7 - 10$ μ.

illuminating source is located a distance z_c from H', which is generally different than the distance z_R of the reference source from the hologram plane during recording.

Figures 8.19a and b show two examples of what has been achieved in practice by Leith and Upatnieks [13]. For these, $\lambda_R = \lambda_c = .633 \ \mu$ so that $\mu = m = 1$. Figure 8.19a is a portion of a fly's wing, taken at $M = 60$. Figure 8.19b is a bar pattern taken at $M = 120$. The spacing between the closest bars is about 8 μ.

8.5   IMAGERY THROUGH DIFFUSING MEDIA [14]

One possible use for holograms might be detection of a spatial signal in the presence of spatial noise—either stationary or time-varying. As an example of the former, consider the simple case illustrated in Fig. 8.20. The object transparency, described by a transmission function $t(x)$, is invisible as viewed from the hologram plane H, because of the presence of the diffusing plate described by the transmission function $d(x)$. It is possible, however, to form an image of the transparency alone by using the conjugate image. To do this, one forms a hologram in the usual way, as indicated in the figure. We suppose that the disturbance at the hologram plane, in one dimension, can be represented by a compound function

$$O(x) = T(x) + D(x) \tag{8.50}$$

where $T(x)$ is the portion of the total object wave due to the transparency and $D(x)$ is the diffuser component, which makes observation of $T(x)$ alone impossible. Addition of a reference wave $R(x)$ yields an exposure (omitting constants)

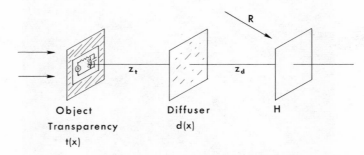

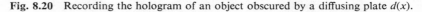

Fig. 8.20   Recording the hologram of an object obscured by a diffusing plate $d(x)$.

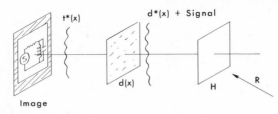

Fig. 8.21 Recovering an image of the originally obscured object by using the diffuser as a complex spatial filter.

$$|H|^2 = |T + D + R|^2$$

$$= |T + D|^2 + |R|^2 + R^*(T + D) + R(T^* + D^*). \qquad (8.51)$$

By illuminating the hologram with a wave $R(x)$ (Fig. 8.21), one of the reconstructed waves (conjugate wave) is

$$R^2(T^* + D^*). \qquad (8.52)$$

By the time this wave has propagated a distance z_d (the distance of the diffuser from the hologram plane during recording), $D^*(x)$ has become $d^*(x)$, that is, the complex conjugate of the original diffuser function $d(x)$. If the identical diffuser $d(x)$ is placed in this position, the wave transmitted by the diffuser becomes $d(x)d^*(x) = d^2 =$ constant, provided $d(x)$ is pure imaginary, that is, a phase disturbance only. Thus the noisy part of the reconstructed wavefront has been filtered out; only the portion corresponding to the transparency remains. By the time the wave (8.52) has propagated a distance z_t, the function $t^*(x)$ is imaged, easily recognizable as the object transparency.

This procedure presents an interesting coding scheme. A recognizable image will result only if the exact diffuser $d(x)$ is available, and it must be relocated in exactly the correct position. If $d(x)$ is not available, $t(x)$ can never be isolated. If the image (conjugate) of the diffuser and the diffuser itself are not precisely coincident, a noisy pattern instead of $t^*(x)$ is produced.

If the noisy medium between the object and hologram is time varying, and thus irreplaceable, the preceding method is of no help. There is still a holographic technique which yields a better image than could be obtained conventionally [15]. Suppose, as indicated in Fig. 8.22, an object is viewed from some plane H, but the time-varying, low spatial frequency phase perturbations of the intervening medium tend to obscure the image. Atmospheric turbulence and rough water would represent such media. If a holo-

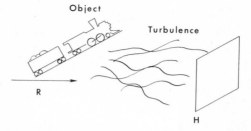

Fig. 8.22 Recording a hologram of an object obscured by a turbulent atmosphere. By keeping the reference beam close to the object beam, the phase difference between the two waves at the recording plane is substantially independent of the turbulence; the hologram is recorded almost as if there were no turbulence.

gram of the object is recorded so that the reference wave traverses substantially the same path as the object wave, it is possible that a usable image can be formed. This is possible because, to a first approximation, the random phase variations introduced onto the object wave are also imposed on the reference wave. Therefore the phase difference between these two waves at the hologram will be largely independent of the intervening medium; the resulting hologram will be substantially the same, regardless of the presence or absence of the perturbing medium.

8.6 THREE-DIMENSIONAL OBSERVATION

One of the most striking aspects of the modern hologram is the three-dimensional image which it is capable of producing. This three-dimensional image indicates that there is a large amount of information contained in a single hologram, certainly much more than is contained in a conventional photograph of the same size. This is especially true when the object in question is many times the depth of field in depth. Because of the many perspectives which are available, the hologram is well suited to display purposes. With a hologram, one can present all of the observable characteristics of a three-dimensional object in a clear and concise manner. Complicated molecular or anatomical structure can be simply presented with a single holographic image, with little chance of error or misinterpretation on the part of the viewer. Such a hologram would take the place of several conventional drawings or photographs. The use of holograms in textbooks would be a great aid to the student in many areas. Holograms made so that they are viewable with a small pen light and a colored filter have already been produced in large quantities and distributed in magazines and books.

The use of holographic images for simulation systems has some draw-backs, such as magnification and/or power for illumination, but these are difficulties easily overcome if the need and usefulness outweigh the costs. Their use as training devices may well prove to be quite advantageous.

8.7   INFORMATION STORAGE

8.7.1   Introduction

The use of holograms as information-storage devices has been one of the most promising right from the start. Initially, we tend to think that a hologram is capable of storing much more information on a two-dimensional medium than a photograph can. This is certainly true if the photograph is merely an image of the information. But there are more suitable ways to photographically store information which make the contest between holography and photography about even. The question of whether or not holography is the best means for storing information has tantalized scientists and engineers. The question has not yet been resolved, but there is still significant effort being applied in this area.

Further, when the third dimension is added—the depth of the holographic recording medium—it becomes clear that holography looks very promising indeed. The gain in storage capacity by utilizing the depth of a photographic emulsion is very real but not nearly as significant as the gains which may be realized with the use of the very thick (of the order of 1 mm or more) photochromic materials.

The proven information capacity of a two-dimensional photographic emulsion is of the order of 10^8 bits/cm², whether the information is in the form of a binary code or microimage. This is for a signal-to-noise (S/N) ratio of about 10. This is the number holography must surpass. On a commercial basis, conventional imaging techniques are used for storage capacities of the order of 10^3 bits/cm² with a S/N of the order of 10^3. For holographic data storage, in two dimensions, we shall see that the capacities are about the same. Holographic data-storage techniques do offer two possible advantages: (*a*) the possibility of the utilization of a three-dimensional recording medium, and (*b*) a large redundancy because of the way in which the data are stored.

8.7.2   Coherent and Incoherent Superposition

Before we make an order-of-magnitude estimate of the storage capacity of a hologram, we should discuss the method of recording. In conventional optical data storage, the information is stored either in the form of a binary code (small patches of density or no density, or possibly several density levels), with the presence or absence of density representing one "bit" of information, or in the form of microimages of the information. With holographic data storage, each bit is stored as a single frequency grating pattern. Each point in object space thus represents a bit of information and this information is stored over an extended area of the holographic recording medium. With the hologram method, each bit can be recorded separately through multiple exposures (incoherent superposition) or, alternatively, all of the bits can be recorded simultaneously (coherent superposition). The information stored in a hologram is read out by reconstructing the plane (or spherical) object wave corresponding to each bit. The strength of the readout signal is proportional to the variance of the amplitude transmittance of the hologram—and this is proportional to the input exposure modulation. We will now show that for coherent superposition, this modulation decreases as $N^{-\frac{1}{2}}$ with increasing N, while for incoherent superposition it decreases as N^{-1}, where N is the number of bits recorded. This means that more flux will be diffracted per bit of information when the bits are recorded coherently than when they are recorded incoherently.

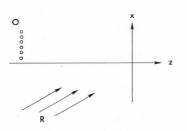

Fig. 8.23 A hypothetical hologram recording system for determining the advantages of coherent vs. incoherent superposition of holograms.

To show this we refer to Fig. 8.23. The object O consists of many discreet points, each representing one bit of information. For coherent superposition, all object points are illuminated simultaneously, so that the light from each object point interferes with the reference beam at the hologram. Also, each object point will interfere with each other point.

For incoherent superposition, each object point interferes with the reference beam sequentially. The reference beam is now attenuated so that the ratio of irradiance at the hologram from a given object point to that of the reference beam is the same as the ratios of irradiance of the total object to reference beam in the case of coherent superposition, so that the total exposure for each case is the same. Since the object points are exposed in sequence, the interference pattern due to an object point and the reference beam will simply add in irradiance to the other interference patterns.

We first assume that the amplitude at the hologram from each of the N object points can be described by $O_m e^{i\varphi_m}$, and that all of the object waves have the same amplitude O_1. The reference wave is described by $R_o e^{ikx}$.

The total irradiance at the hologram plane due to the object is NO_1^2, and the beam balance ratio is

$$B = \frac{NO_1^2}{R_o^2}.$$ (8.53)

When the hologram is recorded, the exposure is

$$E = |O_1 \sum_{n=1}^{N} e^{i\varphi_m} + R_o e^{ikx}|^2$$

$$= NO_1^2 + R_o^2 + 2O_1 R_o \cos(\varphi_1 - kx) + \cdots.$$ (8.54)

This contains a large number of cross terms which appear as cosine functions having various arguments. But we are interested in knowing the modulation of only one such component. This modulation is given by

$$M_c = \frac{2O_1 R_o}{NO_1^2 + R_o^2}$$ (8.55)

where the subscript c refers to the modulation for coherent superposition. If we let $A_t^2 = NO_1^2 =$ the total irradiance from the object, we have

$$M_c = \frac{2A_t R_o}{N^{1/2}(A_t^2 + R_o^2)}.$$ (8.56)

For incoherent superposition, the light from a single object point interferes with the reference beam. The exposure in this case is

$$E_1 = |O_1 e^{i\varphi_1} + N^{1/2} R_o e^{ikx}|^2$$

$$= O_1^2 + \frac{R_o^2}{N} + 2\frac{O_1 R_o}{N^{1/2}} \cos(\varphi_1 - kx).$$ (8.57)

The total exposure received will result from the superposition of N such interference patterns,

$$E = \sum_{n=1}^{N} \left[O_n^2 + \left(\frac{R_o^2}{N}\right)_n + 2\frac{O_1 R_o}{N^{1/2}} \cos(\varphi_n - kx) \right]$$

$$= NO_1^2 + R_o^2 + 2\frac{O_1 R_o}{N^{1/2}} \sum_{n=1}^{N} \cos(\varphi_n - kx).$$ (8.58)

The beam balance ratio for each exposure of the form (8.57) is

$$B = \frac{O_1^2}{R_o^2/N} = \frac{NO_1^2}{R_o^2}$$ (8.59)

which is the same as (8.53), as desired. The modulation for a single interference component is obtained from (8.58) as

$$M_{\text{inc}} = \frac{2O_1 R_o N^{-\frac{1}{2}}}{NO_1^2 + R_o^2} \cdot \tag{8.60}$$

Again letting $A_t^2 = NO_1^2$, we obtain

$$M_{\text{inc}} = \frac{2A_t R_o}{N(A_t^2 + R_o^2)} \cdot \tag{8.61}$$

A comparison of (8.56) and (8.61) shows that for coherent superposition, the modulation varies as $N^{-\frac{1}{2}}$, whereas for incoherent superposition it varies as N^{-1}. Except for nonlinearities of the recording medium, the amount of light flux diffracted from the hologram is proportional to the square of the modulation. Thus for coherent superposition the signal flux will be proportional to N^{-1}.

On the other hand, for incoherent superposition the signal flux varies as N^{-2}, which means that the number of bits which can be recorded in this way is severely limited.

Note that for coherent superposition, each of the N holograms contributes a proportion N^{-1} to the total diffracted flux so that the total diffracted light is independent of N. For incoherent superposition the total diffracted flux goes as N^{-1}.

The foregoing clearly indicates that the best possible way to store information in a hologram is coherently, that is, all of the information is stored simultaneously. This is not always possible, of course, since the input (object) plane may not be large enough to fit all of the information simultaneously. Whenever possible, though, it is desirable to record the information in a single exposure.

8.7.3 Storage Capacity

To obtain an order-of-magnitude estimate of the storage capacity of a hologram, we proceed as follows: Let a diffuse object D be placed a distance f from a lens, with the hologram plane a distance f behind the lens. We

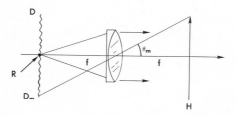

Fig. 8.24 A hologram recording arrangement useful for calculating the information storage capacity of a hologram.

allow a small hole in the center of the object for the reference beam source (Fig. 8.24). We consider the object to consist of a large number of just resolvable points. We now inquire: What is the greatest number N_m of points which can be recorded on a recording medium which can only record spatial frequencies up to $\nu_c l/mm$?

To store the maximum number of points, the object D must be made as large as possible. The relationship between object size and recorded spatial frequency is

$$\nu = \frac{\sin \theta}{\lambda} = \frac{D}{f\lambda} \tag{8.62}$$

so that the maximum allowable object dimension D_m is given by

$$\nu_c = \frac{\sin \theta_m}{\lambda} = \frac{D_m}{f\lambda}. \tag{8.63}$$

The hologram must be at least as large as D_m, and the lens is assumed to be larger than this so that we will not be concerned with the problem of vignetting.

The information stored in the hologram is read out (Fig. 8.25) by illuminating it with a plane wave and forming the image with a lens of focal length f. If the holographic system is diffraction limited, the spot size of the image of a single object point is given by

$$ss = \frac{f\lambda}{H}. \tag{8.64}$$

Therefore the maximum number of resolvable points in the image is

$$N_m = \frac{\text{area of image}}{\text{area of point image}} = \frac{D_m^2 H^2}{f^2 \lambda^2} \tag{8.65}$$

which by (8.63) becomes

$$N_m = \nu_c^2 H^2. \tag{8.66}$$

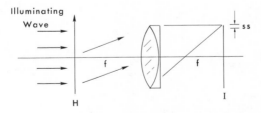

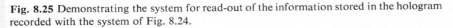

Fig. 8.25 Demonstrating the system for read-out of the information stored in the hologram recorded with the system of Fig. 8.24.

Assuming that we have used a recording medium with the largest available resolving power (ν_c as large as possible), we will want to use this same material for recording the readout in plane I (Fig. 8.25). Thus the material must be capable of resolving the minimum spot size $ss = f\lambda/H$, so that we must have

$$\frac{H}{f\lambda} \leq \nu_c = \frac{D_m}{f\lambda}. \tag{8.67}$$

Hence we should have $H = D_m$. If $H > D_m$, then we would be able to store more information but would lose on the readout since all of the signal points would not be resolved. If we make $H < D_m$, we will unnecessarily restrict N_m, since the minimum resolvable spot size will be larger than necessary.

The number N_m given by (8.66) is the maximum number of points (bits of information) we can store on a single two-dimensional hologram of dimension H. We must now determine the signal-to-noise ratio S/N. All recording media containing a stored signal will scatter some of the incident light during readout. It is this scattered light which we shall call the noise of the process. In holography, readout is often accomplished by illuminating the hologram with laser light, therefore the noise is coherent with the signal. This fact somewhat complicates the problem of defining a meaningful signal-to-noise ratio, but for our present purpose it will be sufficient to define S/N as the ratio of the signal flux entering the cone subtending the solid angle $\Omega_s = \pi\lambda^2/4H^2$ to the scattered flux within the same solid angle. The solid angle Ω_s is defined in Fig. 8.26; it is just the solid angle subtended by the image spot at the hologram.

We first compute the noise. The hologram is assumed to scatter some fraction R of the incident flux F_o into π steradians:

$$R = \frac{\text{scattered flux}}{\text{incident flux}}. \tag{8.68}$$

The flux scattered into the same solid angle as the signal is just

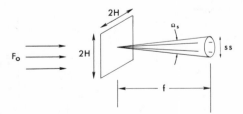

Fig. 8.26 Defining the solid angle Ω_s—the solid angle subtended by the area of a point image at the hologram.

$$N = RF_o \frac{\Omega_s}{\pi} = RF_o \frac{\lambda^2}{4H^2}. \tag{8.69}$$

The signal flux is also proportional to F_o and we take as the signal the amount of flux diffracted by the hologram into a single, just resolvable spot. Hence we can write the signal as

$$S = \beta \frac{F_o}{N_m} \tag{8.70}$$

where β is the fraction of the incident flux F_o diffracted into either the primary or conjugate holographic image. The fraction contributing to the signal must then be just $1/N_m$ of this flux. This fraction β can be calculated from the known parameters as shown in Chapter 5. In particular, (6.77) shows that the ratio of diffracted flux to object flux is

$$K_i = 2g'^2(E_o) \frac{E_o^2}{(1 + B)^2} \tag{8.71}$$

where $g'(E_o)$ is the slope of the t-E characteristic curve of the recording medium evaluated at E_o. The average exposure E_o is just the total average irradiance at the hologram plane, $\langle O^2(x) \rangle + R^2$, since we assumed unit exposure time in deriving (6.77). The beam balance ratio B is $\langle O^2(x) \rangle / R^2$. Since (6.77) was derived assuming that the processed hologram was illuminated with irradiance $C^2 = R^2$, we have by definition

$$K_t = \frac{2\beta R^2}{\langle O^2(x) \rangle} \tag{8.72}$$

so that

$$\beta = g'^2(E_o) \frac{BE_o^2}{(1 + B)^2}. \tag{8.73}$$

The factor of 2 in (8.72) comes from the fact that K_i refers to the total flux diffracted into both images, and we are here interested in only the fraction diffracted into one image. By substituting (8.73) into (8.70) we obtain

$$S = g'^2(E_o) \frac{F_o B E_o^2}{N_m(1 + B)^2}. \tag{8.74}$$

Using (8.69) we arrive at the signal-to-noise ratio

$$\frac{S}{N} = g'^2(E_o) \frac{4H^2 B E_o^2}{R N_m \lambda^2 (1 + B)^2}. \tag{8.75}$$

By using (8.66) this can be written

$$\frac{S}{N} = 4g'^2(E_o) \frac{BE_o{}^2}{R\nu_c{}^2\lambda^2(1 + B)^2} . \tag{8.76}$$

A reasonable estimate of the information capacity N_m/H^2 and the S/N ratio can be determined as follows: We suppose that $H = 10$ cm. and $\nu_c = 10^4$ l/cm (1000 l/mm). Equation 8.66 then gives $N_m = 10^{10}$ as the maximum number of point object holograms which may be stored. The information capacity is therefore some 10^8 bits/cm^2, which is the same capacity as with conventional storage techniques. To arrive at a figure for the S/N of such a holographic storage system, we need an estimate of the R of (8.68) and β of (8.73). To make these estimates, we will use actual experimental results obtained with a photographic emulsion, Kodak Spectroscopic Plate, Type 649F, a commonly used holographic recording medium. An easy way to estimate R is by using the Callier Q coefficient. The Callier Q is defined as

$$Q = \frac{\text{specular density}}{\text{diffuse density}} . \tag{8.77}$$

The diffuse density D_d is given by

$$D_d = \log_{10} \frac{F_o}{F_T} \tag{8.78}$$

where F_o is the incident flux and F_T is the total flux transmitted into π steradians. The specular density D_s is

$$D_s = \log_{10} \frac{F_o}{F_s} \tag{8.79}$$

where F_s is the flux transmitted into a vanishingly small solid angle. Since D_d includes all of the scattered flux, we always have $D_d \leq D_s$ or

$$Q \geqslant 1. \tag{8.80}$$

We have

$$Q = \frac{D_s}{D_d} = \frac{\log_{10}(F_o/F_s)}{\log_{10}(F_o/F_T)} \tag{8.81}$$

so that

$$\frac{F_T}{F_o} = \left(\frac{F_s}{F_o}\right)^{1/Q} . \tag{8.82}$$

Since F_T is the total flux transmitted into π steradians, it can be divided into specular and scattered components:

$$\frac{F_T}{F_o} = \frac{F_{\text{scat}} + F_s}{F_o} . \tag{8.83}$$

But $R = F_{\text{scat}}/F_o$ so that

$$R = \left(\frac{F_s}{F_o}\right)^{1/Q} - \frac{F_s}{F_o}. \tag{8.84}$$

For Kodak Spectroscopic Plate, Type 649F, processed to a specular transmittance of 0.25, the Callier Q is approximately 1.1 so that

$$R = 0.033. \tag{8.85}$$

A value for β comes directly from the curve of Fig. 6.20. Here we see that for the same plate, exposed with a beam balance ratio of 1 and processed to a transmittance of about 0.25,

$$\beta = \frac{g'^2(E_o)BE_o^2}{(1 + B)^2} \cong 0.02. \tag{8.86}$$

Using these numbers in (8.76) yields a signal-to-noise ratio

$$\frac{S}{N} = \frac{\pi\beta}{R\nu_c^2\lambda^2} \approx 5. \tag{8.87}$$

This S/N is somewhat low, but it is based on conservative estimates. Also, there are various schemes, for example, bleaching, which yield somewhat larger values of β.

Thus far we have not surpassed the storage capabilities of photographic film used in the conventional manner—however, there is still a third dimension to be considered. Recall that in Chapter 4 we discussed the effects of emulsion thickness on the properties of the hologram. There we found that diffraction is governed by the Bragg condition—both the grating equation and the law of reflection must hold simultaneously with regard to the directions of the illuminating and diffracted waves. In particular, it was found that for thick recording media, only a small misorientation of the illuminating beam is required to extinguish the image. For recording media of the order of 1 mm thick, such as the photochromics, this angular orientation tolerance may be as small as a few seconds of arc. Just how many holograms can be stored on a single plate in this manner is difficult to determine. The signal-to-noise ratio (8.87) will certainly fall off with each additional hologram, but just how fast the S/N declines has not been determined. In a noiseless system, with a cut-off frequency of 10^3 l/mm and an orientation tolerance of 10 seconds of arc, the ultimate storage capacity on a 10×10 cm plate might be some 3.6×10^{14} bits of information:

$$\frac{\text{angular range}}{\text{orientation tolerance}} \times N_m = \frac{100^{\circ}}{\frac{1}{360}^{\circ}} \times 10^{10} = 3.6 \times 10^{14}.$$

Even if only 10 maximum capacity holograms could be stored on a single plate, with a reasonable S/N level, the gain over existing information capacity limits is some $10\times$, a not insignificant factor.

8.8   ULTRASONIC HOLOGRAPHY

Ultrasonic, or acoustical holography is a new variant on the general method of holography. The recording and readout techniques so far applied have been rather complex and cumbersome and the results have been of very poor quality compared to what has been achieved with light. Effort in this area is just beginning, however, and there is little doubt that simpler and more reliable recording techniques will evolve, with subsequent improvement in results. The vast number of interesting applications for acoustical holography warrant the already sizeable effort in this area, and certainly some of these possible applications deserve mention in this section.

Most of the principles pertaining to light holography which have been discussed in the preceding chapters also apply to ultrasonic holography. Ultrasonic holography is a two-step process wherein the diffraction pattern of an object irradiated with ultrasonic waves is interfered with a mutually coherent reference wave. The resulting spatial irradiance distribution is recorded in some manner. Next the acoustical hologram is illuminated with a beam of light. The resultant diffraction from the hologram provides a generally reduced object wave, which can be used to form a three-dimensional visual image of the object.

The geometrical arrangements for making ultrasonic holograms are quite straightforward. One possible arrangement is shown in Fig. 8.27. The ultrasonic transducer S irradiates the object O and the resulting scattered wave is reflected to the water surface by means of a plane mirror. A portion of the acoustic wave is intercepted by the plane mirror M and directed toward the water surface as a reference wave. At the water surface the acoustical object and reference waves interfere in a manner characteristic of the object O. This interference pattern appears as ripples on the surface of the water. The problem of suitably recording this pattern has been approached in several ways, but none to date has been very satisfactory.

One method which has been used requires photographing the ripple pattern on a water surface directly by means of a schlieren system [16]. Other possible

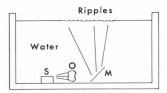

Fig. 8.27 Schematic representation of a possible ultrasonic hologram recording arrangement. The source of ultrasound S illuminates an object O and a mirror M directly. These two waves interfere at the water surface, forming the ripples. The ripple pattern is recorded in some manner and this recording constitutes the hologram.

schemes include recording the waves on the water surface directly with a pressure-sensitive recording medium [17], scanning the acoustic field with an electrical transducer [18, 19, 20], or by driving the source hard enough to induce cavitation on a water surface and using the resulting localized turbulence to process photographic film [21]. With ultrasonics it is not a necessary requirement for coherence that both object and reference beams originate from the same source. One can just as well use separate drivers for the object and reference beams, as long as they are mutually coherent.

The final image formation is achieved with more or less conventional means. The acoustical hologram, assuming it has been recorded in the form of a transparency, is illuminated with a collimated monochromatic beam of light. Since this wavelength will generally be much smaller than the acoustical wavelength used to record the hologram, good imagery can be achieved only when the hologram is reduced in size by a fraction equal to the ratio of the optical to the acoustical wavelengths. The images are then formed in the usual manner.

The principal feature of ultrasonic holography is its ability to form three-dimensional images of objects which are opaque to light waves but transmit ultrasonic waves. Thus the interior of solid parts might be examined optically. Because of this feature, underwater and underground surveillance and exploration may be possible. Also, acoustical holography may find important applications in the medical field. Parts of the human body, for example, which are transparent to sound waves, may be rendered visible by means of ultrasonic holography, with the image presented in a much more useful and easily interpreted manner than the simple shadowgrams now produced with x-rays.

REFERENCES

[1] L. H. Tanner, *J. Sci. Instr.*, **43**, 81 (1966).

[2] K. A. Haines and B. P. Hildebrand, *Appl. Opt.* **5**, 595 (1966).

[3] L. O. Heflinger, R. F. Wuerker, and R. E. Brooks, *J. Appl. Phys.*, **37**, 642 (1966).

[4] R. E. Brooks, L. O. Heflinger, and R. F. Wuerker, *IEEE J. Quantum Electronics*, **QE-2**, 275 (1966).

[5] R. E. Brooks, L. O. Heflinger, and R. F. Wuerker, *Appl. Phys. Letters*, **7**, 248 (1965).

[6] R. E. Brooks, R. F. Wuerker, L. O. Heflinger, and C. Knox, Paper presented at the International Colloquium on Gasdynamics of Explosions, Brussels, Belgium, September 20, 1967.

[7] J. M. Burch, A. E. Ennos, and R. J. Wilton, *Nature*, **209**, 1015 (1966).

[8] R. L. Powell and K. A. Stetson, *J. Opt. Soc. Am.*, **55**, 1593 (1965).

[9] K. Haines and B. P. Hildebrand, *Phys. Letters*, **19**, 10 (1965).

[10] J. Upatnieks, A. Vander Lugt, and E. Leith, *Appl. Opt.*, **5**, 589 (1966).

[11] A. Vander Lugt, *IEEE Trans. Information Theory*, **IT-10**, 139 (1964).

[12] C. B. Burckhardt, *Appl. Opt.*, **6**, 1359 (1967).

[13] E. N. Leith and J. Upatnieks, *J. Opt. Soc. Am.*, **55**, 569 (1965).

[14] E. N. Leith and J. Upatnieks, *J. Opt. Soc. Am.*, **56**, 523 (1966).

[15] J. W. Goodman, *Appl. Phys. Letters*, **8**, 311 (1966).

[16] R. K. Mueller and N. K. Sheridon, *Appl. Phys. Letters*, **9**, 328 (1966).

[17] J. D. Young and J. E. Wolfe, *Appl. Phys. Letters*, **11**, 294 (1967).

[18] K. Preston, Jr. and J. L. Kreuzer, *Appl. Phys. Letters*, **10**, 150 (1967).

[19] A. F. Metherall, H. M. A. El Sum, J. J. Dreher, and L. Larmore, *Appl. Phys. Letters*, **10**, 277 (1967).

[20] G. A. Massey, *Proc. IEEE* (*Letters*), **55**, 1115 (1967).

[21] P. Greguss, *J. Phot. Sci.*, **14**, 329 (1966).

Appendix. Fourier Transforms with Lenses

A.0 INTRODUCTION

Several authors [1, 2, 3] have shown that within certain limitations (not always explicitly stated) a lens forms, in its back focal plane, the Fourier transform of the field distribution in the front focal plane. Since this property of a lens is used so often throughout this book, we present here our own derivation, so that symbols and phase conventions will be well defined. We will treat only the one-dimensional case.

A.1 ANALYSIS

The starting point in any diffraction problem is Huygens' principle, which, in one dimension (cylindrical waves) may be written

$$g(x_L) = \left(\frac{-i}{4\lambda}\right)^{\frac{1}{2}} \int_{-\infty}^{\infty} F(x_o)[1 + \cos\theta] \frac{e^{ikr}}{(r)^{\frac{1}{2}}} dx_o. \tag{A.1}$$

With this formulation, we assume that each elementary strip of a wavefront $F(x_o)$ acts as a line source of cylindrical waves that lead the parent wave in phase by $\pi/4$. The amplitude contributed by this element at some distant point at the proper later time is the wave amplitude of the parent wave times the factor

$$\frac{1 + \cos\theta}{2(\lambda r)^{\frac{1}{2}}}. \tag{A.2}$$

The factor $\frac{1}{2}(1 + \cos\theta)$ is the obliquity factor; θ is the angle between the line from dx_o to x_L and the normal to dx_o (see Fig. A.1). The length r is the distance between the strip dx_o and the point x_L on the lens at which we wish to know the field.

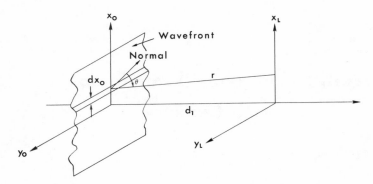

Fig. A.1   Illustrating the notation and meaning of (A.1).

If we restrict the analysis to small fields, for example, $\theta \leq 10°$, then there is only a negligible error in writing

$$\tfrac{1}{2}(1 + \cos \theta) \approx 1, \tag{A.3}$$

and under this assumption we may also bring the factor $r^{-\frac{1}{2}}$ outside of the integral A.1, since this will not change significantly as dx_o moves over the wavefront. Actually, there is a much more stringent requirement on θ imposed by the approximations which follow, so the approximation A.3 will always be quite good.

With these approximations, the statement of Huygens' principle becomes

$$g(x_L) = \left(\frac{-i}{\lambda d_1}\right)^{\frac{1}{2}} \int_{-\infty}^{\infty} F(x_o)e^{ikr}\, dx_o \tag{A.4}$$

where we have used $r^{-\frac{1}{2}} \cong d_1^{-\frac{1}{2}}$. The factor e^{ikr} describes the phase of the secondary wavelet. We now apply this formulation to the situation shown in Fig. A.2. The x_o, y_o plane will be taken as the object plane and we suppose that we can describe the disturbance in this plane as $F(x_o)$. The x_L, y_L plane is the plane in which the lens is located. Our ultimate goal is to find the disturbance in the x-y plane and to show that this disturbance looks very much like the Fourier transform of $F(x_o)$.

We first write

$$r = [(x_o - x_L)^2 + d_1^2]^{\frac{1}{2}} \cong \pm d_1 \pm \frac{(x_o - x_L)^2}{2d_1}. \tag{A.5}$$

Since d_1 is a positive number, r will always be positive if we choose the upper signs of (A.5). Phase is advancing for the wave traveling left-to-right (positive z direction) and for $r \geq 0$ we simply write this as e^{ikr}. It is the

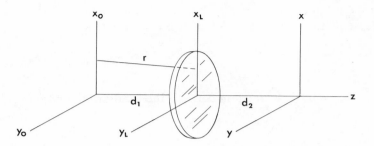

Fig. A.2   Illustrating the notation and geometry for determining the effect of a lens on a wavefront.

approximation (A.5) which restricts the value of θ to only a few degrees (see [2] for a more detailed discussion of this approximation). We thus have for the disturbance just in front of the lens

$$g(x_L) = \left(\frac{-i}{\lambda d}\right)^{1/2} \int_{-\infty}^{\infty} F(x_o)e^{ikd_1}e^{ik[(x_o-x_L)^2/2d_1]}\, dx_o. \tag{A.6}$$

We know that the action of the lens must be such as to render a wave from a point source on the axis at $x_o = 0$ plane when $d_1 = f$, f being the focal length of the lens. To do this the lens must retard the phase of the wave more at the center than at the edges. The phase-retardation factor can easily be derived for the equiconvex lens of Fig. A.3. When a plane wave is incident from the left, the optical path difference between the axial ray and a ray incident at a height x is just $2S(n-1)$ for a lens with index n. The sag S is given by

$$S \cong \frac{x_L^2}{2R} \tag{A.7}$$

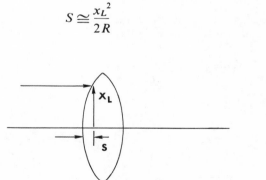

Fig. A.3   The optical path difference between an axial ray and a ray at a height x_L for an equiconvex lens.

with the usual assumption that $R \gg S$, where R is the radius of curvature of the (equal) lens surfaces. The optical path difference is

$$\frac{(n-1)x_L^2}{R}. \tag{A.8}$$

For a simple lens, the relation between the focal length and radius is

$$f = \frac{R}{2(n-1)} \tag{A.9}$$

so that the optical path difference may be written in terms of the focal length as

$$\frac{x_L^2}{2f}. \tag{A.10}$$

Thus the phase factor introduced by the lens of Fig. A.2 is just

$$e^{-ik(x_L^2/2f)} \tag{A.11}$$

and the disturbance (A.6), after passing through the lens, becomes

$$\psi(x_L) = g(x_L)e^{-ik(x_L^2/2f)}.$$

By repeating the application of Huygens' principle in the form (A.6) we can now calculate the disturbance in the x, y plane of Fig. A.2:

$$O(x) = \left(\frac{-i}{\lambda d_2}\right)^{1/2} \int_{-\infty}^{\infty} \psi(x_L)e^{ikd_2} \exp\left[ik\frac{(x_L-x)^2}{2d_2}\right] dx_L$$

$$= \frac{-ie^{ik(d_1+d_2)}}{\lambda(d_1d_2)^{1/2}} \iint_{-\infty}^{\infty} F(x_o) \exp\left[ik\frac{(x_o-x_L)^2}{2d_1}\right] \exp\left[ik\frac{(x_L-x)^2}{2d_2}\right]$$

$$\times \exp\left(-ik\frac{x_L^2}{2f}\right) dx_o\, dx_L$$

$$= \frac{-ie^{ik(d_1+d_2)}}{\lambda(d_1d_2)^{1/2}} \exp\left(ik\frac{x^2}{2d_2}\right) \iint_{-\infty}^{\infty} F(x_o) \exp\left(ik\frac{x_o^2}{2d_1}\right)$$

$$\times \exp\left[i\frac{k}{2}x_L^2\left(\frac{1}{d_1}+\frac{1}{d_2}-\frac{1}{f}\right)\right] \exp\left[-ikx_L\left(\frac{x_o}{d_1}+\frac{x}{d_2}\right)\right] dx_o\, dx_L.$$

$$\tag{A.12}$$

But

$$\int_{-\infty}^{\infty} \exp\left[i\frac{k}{2}x_L{}^2\left(\frac{1}{d_1} + \frac{1}{d_2} - \frac{1}{f}\right)\right] \exp\left[-ikx_L\left(\frac{x_o}{d_1} + \frac{x}{d_2}\right)\right] dx_L$$

$$= \left[\frac{\lambda}{-i(1/d_1 + 1/d_2 - 1/f)}\right]^{1/2} \exp\left[\frac{-i(k/2)(x_o/d_1 + x/d_2)}{(1/d_1 + 1/d_2 - 1/f)}\right], \quad \text{(A.13)}$$

so

$$O(x) = \left(\frac{-i}{\lambda}\right)^{1/2} e^{ik(d_1+d_2)} \exp\left(ik\frac{x^2}{2d_2}\right) \exp\left[-i\frac{k}{2}\frac{x^2(d_1/d_2)f}{f(d_1 + d_2) - d_1d_2}\right]$$

$$\times \int_{-\infty}^{\infty} F(x_o) \exp\left(ik\frac{x_o{}^2}{2d_1}\right) \exp\left[-i\frac{k}{2}\frac{x_o{}^2(d_2/d_1)f}{f(d_1 + d_2) - d_1d_2}\right]$$

$$\times \exp\left[-ik\frac{xx_of}{f(d_1 + d_2) - d_1d_2}\right] dx_o. \quad \text{(A.14)}$$

We see that the quadratic exponentials under the integral will cancel if we choose d_1 and d_2 so that

$$\frac{(d_2/d_1)f}{f(d_1 + d_2) - d_1d_2} = \frac{1}{d_1} \quad \text{(A.15)}$$

which in turn implies

$$d_2 f = f(d_1 + d_2) - d_1d_2. \quad \text{(A.16)}$$

Also, to make (A.14) look like a Fourier transform, we must cancel the quadratic exponentials in x in front of the integral, or

$$d_1 f = f(d_1 + d_2) - d_1d_2. \quad \text{(A.17)}$$

Equations A.16 and A.17 can hold simultaneously only for $d_1 = d_2$, and this implies $d_1 = d_2 = f$. Thus (A.14) becomes

$$O(x) = \left(\frac{-i}{\lambda f}\right)^{1/2} e^{2ikf} \int_{-\infty}^{\infty} F(x_o)e^{-i(k/f)xx_o}dx_o. \quad \text{(A.18)}$$

It is usual practice to neglect the constant phase factor e^{2ikf} since it is not of any importance. Even without this factor, however, (A.18) is still not in the form of a Fourier transform, so it is not necessary to neglect it. To obtain a true Fourier relationship, we first define a spatial frequency ν as

$$\nu \equiv \frac{x}{f\lambda}. \quad \text{(A.19)}$$

Then (A.18) becomes

$$O(\nu\lambda f) = \left(\frac{-i}{\lambda f}\right)^{\frac{1}{2}} e^{2ikf} \int_{-\infty}^{\infty} F(x_o)e^{-2\pi i\nu x_o} \, dx_o. \tag{A.20}$$

By defining a new function

$$V(\nu) \equiv \left(\frac{-i}{\lambda f}\right)^{-\frac{1}{2}} e^{-2ikf} O(\nu\lambda f), \tag{A.21}$$

(A.20) becomes

$$V(\nu) = \int_{-\infty}^{\infty} F(x_o)e^{-2\pi i\nu x_o} \, dx_o, \tag{A.22}$$

which is the desired Fourier transform relationship.

REFERENCES

[1] J. E. Rhodes, Jr., *Am. J. Phys.*, **21**, 337 (1953).
[2] E. N. Leith, A. Kozma, and C. J. Palermo, Notes from Int. to Opt. Data Processing, Eng. Summer Conferences, University of Michigan, Ann Arbor, Michigan, 1966.
[3] K. Preston, Jr., *Optical and Electro-Optical Information Processing* (Symposium Proceedings, Boston, 1964), J. T. Tippett, D. A. Berkowitz, L. C. Clapp, C. J. Koester, and A. Vanderburgh, Jr., eds., Mass. Inst. of Tech. Press, Cambridge, 1965, p. 59.

Index